SUSTAINING

THE FOREST,

THE PEOPLE,

AND

THE SPIRIT

SUSTAINING THE FOREST, THE PEOPLE, AND THE SPIRIT

Thomas Davis

STATE UNIVERSITY OF
NEW YORK PRESS

Published by
State University of New York Press

© 2000 State University of New York

For information, address the State University of New York Press,
State University Plaza, Albany, NY 12246

Marketing by Anne Valentine • Production by Bernadine Dawes

Library of Congress Cataloging-in-Publication Data

Davis, Thomas, 1946–
        Sustaining the forest, the people, and the spirit / Thomas Davis.
              p.   cm.
        Includes bibliographical references.
        ISBN 0-7914-4415-5 (hc : alk. paper). — ISBN 0-7914-4416-3 (pb :
alk. paper)
              1. Menominee Indians. 2. Sustainable forestry—Wisconsin.
        3. Sustainable development. I. Title.

        E99-M44D38 2000
        333.75′15′0899730775—dc21   99-13229
                                                                          CI

1 2 3 4 5 6 7 8 9 10

Table of Contents

Acknowledgments

The major debt I owe in the writing of this book is to my wife, Ethel, and son, Kevin. How families ever put up with writers is beyond rational understanding. My two daughters, Sonja Bingen and Mary Wood, have provided encouragement.

Equal debts are owed to Dr. Michael Kraft and Dr. John Stoll, who have read through multiple drafts, making corrections and helping me to focus the text. Dr. Verna Fowler, President of the College of the Menominee Nation, has not only responded to my questions about the Menominee, but allowed me time from my job to conduct research.

The best insights into Menominee sustainable development were provided by Menominee like Larry Waukau, Dr. Fowler, Glenn Miller, Marshall Pecore, and Shirley Daly, or colleagues like Keith Milner, Paula Huff, or Dr. Denise Scheberle.

Tables and Figures

1

Sustainable Development Visions and the Menominee Indian Tribe of Wisconsin

IN 1969 CHARLIE FRECHETTE, a Menominee Indian and the vice president of Menominee Enterprises, Inc., quoted an unnamed tribal leader as saying,

> [S]tart with the rising sun and work toward the setting sun, but take only the mature trees, the sick trees, and the trees that have fallen. When you reach the end of the reservation, turn and cut from the setting sun to the rising sun, and the trees will last forever. (Spindler and Spindler 1971, 201)

When this statement was made, the Menominee Indian Tribe was in the midst of the termination experiment by the United States government. Public Law 83-399, known on the Menominee Indian Reservation as the Termination Act, had ended federal responsibility for the Menominee as called for under the treaties signed by the tribe with the U.S. government during a period ranging from 1817 to 1848. In the official language of the Termination Act,

> [T]he responsibility of the United States to furnish all such supervision and services to the tribe and to members thereof because of their status as Indians, shall cease on December 31, 1960 or on such earlier date as may be agreed upon by the tribe and by the secretary [of the Interior]. (PL 83-399, 1954)

In the language of those Menominee who fought to replace termination with a new federal policy that they called restoration,

> Thanks to termination, the tribal treasury has been long since virtually depleted; the lumbering operation teeters on insolvency; new

1

economic developments bring no direct benefit to the living standard
of the community but only keep the tribal corporation [Menominee
Tribal Enterprises] afloat; the hospital is closed, with no medical care
available to the community; the utility companies have been sold;
and hundreds of Menominee have been forced to leave home to find
employment, many only to trade rural poverty for urban slums.
(National Committee to Save the Menominee People and Forests,
1972, 11)

Clearly, Mr. Frechette's statement in 1969 was a remarkable one
for a member of a devastated Indian tribe to make. The Menominee,
in little over a decade, had gone from being one of the wealthier
Indian tribes in the nation, with a cash balance of $9,960,895 invested
with the U.S. Treasury, to one of the most impoverished. The tribe
had just a few years earlier owned its own hospital, had plenty of
work for most of its able-bodied men, ran its own well-functioning
court system, and owned its own utility companies. In 1969 the situation declared by the National Committee to Save the Menominee
People and Forests in 1972 was already true. The Menominee termination experiment had exacted a terrible toll on the Menominee people.

Yet, in 1969 an obvious economic remedy existed for the
Menominee plight. Of the 233,902 acres of unallotted land along the
upper Wolf River in northeastern Wisconsin that made up the reservation prior to termination, approximately 95 percent was still
forested land owned and managed by Menominee Tribal Enterprises.
Although a thicket of federal and State of Wisconsin legislation stood
between the ability to clear-cut these reserves, a consummated sale of
timber harvested wholesale, and the actual decision to clear-cut, the
value of the lumber in these forests was extraordinary. In 1961 the
Menominee forest represented 10 percent of the total standing timber
in the State of Wisconsin and was the largest hardwood forest in the
state. The fair market value of the lumber, if it had been harvested in
1961, was set at 38 million dollars by the Supreme Court of the United
States as the result of a case, *Menominee Tribe of Indians v. The United
States of America*, filed by the tribe in 1981 (Supreme Court of the
United States 1984, 175a). Some appraisers at the time believed that
the Court's valuation was understated. Menominee land, of course,
had a substantial additional value beyond the logs that could be harvested, especially the land shoring lakes, streams, or rivers—all three
of which the Menominee have in abundance. Recreation land was
becoming increasingly valuable in 1960s America, and the Menominee

had a lot of land that possessed the values prized for recreation. During the termination period, this fact would have enormous consequences for the tribe and its future.

Ending Menominee poverty could have been relatively simple. All the Menominee had to do was to convince federal and state governments the termination policy had been what it was—a disaster—and convince the key players from those governments that the only way the Menominee had out of the dilemma they had been forced into by the U.S. Congress was to allow them the freedom to join the free market economy enjoyed and supported overwhelmingly by the Menominee's fellow citizens. Fewer than three thousand Menominee, men, women, and children lived in Menominee County, the political entity established conterminous with reservation boundaries after termination had taken effect in 1961. By selling their forests and land at any point during the termination period, which did not end until passage of the Menominee Restoration Act in 1972, and then carefully investing the resulting capital sums, the Menominee could have taken a large step toward lifting themselves out of poverty for both present and future times.

Still, that was not the Menominee way. The Menominee way was identified by Charlie Frechette when he quoted a Menominee leader: "and the trees will last forever." The Menominee did not object to harvesting the mature, the sick, and the fallen trees, but when all was said, the trees had to last forever. The beauty and substance of the forest, with its white pine canopies, its lake shores crowned with maple, white pine, red pine, and birch, the forested banks of the wild, powerful Wolf River, had to be preserved forever. There was no question about that. The Menominee people might live in poverty, but living in poverty was better than living in a landscape denuded of forest and made to look uncared for and scrubby. In the long term the Menominee were a woodland people, and in the end they would be cared for by the forest.

This attitude, the decisions the Menominee have made since 1908 and the passage of the LaFollette Act that allowed the Menominee a forest harvest of twenty million board feet per year, through the present, constitute a significant environmental record with international consequences. Just how significant this record is can be partially measured by a recent reception held at the United Nations in New York where four indigenous people—the Menominee of Wisconsin, the Quichua from the Amazon forests, the Kuna of Panama, and the Hoopa of British Columbia, Canada—were all honored for their contributions to the world's environment (Shawano Leader, 21 April 1995, 1). In February of 1996 the Menominee were again honored when Vice President Al Gore awarded the Menominee a Presidential Award from the President's Council on Sustainable Development.

The Menominee's contribution, according to Larry Waukau, the current President of Menominee Tribal Enterprises, the company formed after passage of the Restoration Act to succeed Menominee Enterprises, Inc., was the tribe's land ethic and management philosophy that

> has always contained the three elements of a sustainable system. First he [the forest] must be sustainable for future generations. Second, the forest must be cared for properly to provide for the needs of the people. And third, we keep all the pieces of the forest to maintain diversity. (*Shawano Leader* 1995, 1)

Another measure of this significance is the number of professional foresters, environmentalists, university teachers, political leaders, and other visitors who come to see the Menominee forest every year. In 1994 just under 2,000 visitors from all of the world's major continents visited the forest. The count for 1995 and 1996 exceeded that for 1994 (Waukau 1994).

One motive these visitors have for coming to the Menominee homeland is to see the forest and to talk to the foresters and Menominee leaders responsible for preserving the forest, but an equally important motive is the suspicion that the Menominee know something about sustainability and sustainable development the rest of the world does not know. Indigenous people are mentioned as a primary source of information about sustainable development in much of the sustainable development literature, ranging from the Brundtland report (The World Commission on Environment and Development, 1988) to the philosophical, idealistic ruminations of a writer like Thomas Berry (Berry 1988). What the Menominee have achieved with their forest has the feel of what scholars and environmentalists mean by sustainable development. Here is a people, paraphrasing Larry Waukau's words, who have sustained the forest that has been in their care for close to a century and a half. They have worked to keep the forest's diversity. But, at the same time, they have also developed and used the reservation's forest and water resources to properly provide for the needs of the Menominee people.

Certainly the idea of sustainable development has become

increasingly important to the contemporary world. New books on sustainable development have been appearing with increasing rapidity since the United Nations Conference on Environment and Development (the Earth Summit) held in Rio de Janeiro, Brazil, in June of 1992, and the number of articles appearing in professional journals has been expanding at what seems to be an exponential rate.

The link between the way the Menominee manage their forest and the idea of sustainable development is largely assumed throughout the reservation, especially among the reservation's leadership. The College of the Menominee Nation, a tribally founded land grant institution of higher education that has a reservation-based campus, has even established a Sustainable Development Institute that includes many of the most important Menominee leaders on its Advisory Council. The primary purposes of the Institute are to establish a Menominee model of sustainable development and to conduct research that helps the Menominee people use their resource base in a more sustainable manner. Reservation leaders have also not been shy about telling their story in ways that garner the maximum amount of exposure. This is indicated by the application submitted by the Reservation to the President's Council on Sustainable Development for the purpose of receiving President's Award recognition for their sustainable development efforts.

One of this book's burdens is to clearly establish the link between Menominee management and sustainable development, and then to examine both its constituent parts and how it relates to the larger discussion cataloged in the library of literature on sustainable development. The hypothesis is that a careful study of the Menominee record can lead to a better understanding of processes that might lead to a sustainable world. The most significant question this book avoids is whether or not there is wisdom in the pursuit of sustainable development policies in today's world. The literature on the environmental crisis is so massive, and the existence of a state of crisis is so well documented elsewhere, that an assumption is made that all human beings should be seriously studying and thinking about how we, collective mankind, can come into a better balance with the earth's ecological systems. The reason the Menominee example is so important is that it holds the promise of providing a series of lessons about how we might create a sustainable world that can prosper into the future.

The basic question this book pursues is: Can lessons be learned from the Menominee model of sustainable development that can be used to help establish a general theory of sustainable development? Secondary questions are: What are the elements of the Menominee sustainable development model? How significant are the Menominee sustainable development accomplishments? What role have history, culture, spirituality, technology, science, and political structures played in the development of the Menominee sustainable development model? Are there proposals developed in the past or proposals in the current sustainable development or sustainable environmental economics literature that are supported by the Menominee sustainable development experience? What elements of the model are applicable to societal structures larger than those of the Menominee? Does the Menominee model provide clues about the best way to pursue economic development if the goal is to use the world's resources in a sustainable manner? These questions are addressed sequentially in the chapters of this book.

The Task of Construction of a Menominee Model of Sustainable Development

The task of constructing a Menominee model of sustainable development is not an easy one. There are a number of reasons for this. First, the range of material that must be covered is vast. Historical, cultural, religious, economic, bicultural, and anthropological aspects affect the Menominee story along with scientific and technological aspects. The Menominee record itself is spread through primary historical documents, the work of anthropologists, the writings of linguists, newspaper stories, Native American literature, court documents, business documents, treaties between the tribe and the U.S. government, State of Wisconsin legislation, tribal legislation, U.S. legislation, scientific literature, social science literature, and the words and feelings of Menominee themselves. Fortunately, a wealth of material on the Menominee, ranging back to the earliest days of contact with French explorers and Jesuit missionaries, exists. First-rate anthropologists such as Walter J. Hoffman, Alanson Skinner, Felix M. Keesing, George and Louise Spindler, and the linguist Leonard Bloomfield have written book-length studies, and a number of historians, including David Wrone, David Beck, Patricia Ourada, Verne F. Ray, and Nicholas Peroff, have published extensive works on the tribe.

These documents and studies show that the Menominee live in an exceptionally complex, acculturating society. Although considerable tribal cohesiveness exists, at the same time, religious, cultural, economic, and attitudinal differences have resulted from three centuries of contacts with European and American culture. Ideas and attitudes that have led the Menominee to their current understanding and adherence to sustainable development have their roots in both the tribe's cohesion and the acculturative strategies it has used to adapt to a succession of non-Indian societies, starting with the French, then changing to the English, and finally ending up with the American. Both history and religion have played major roles in the creation of the reservation's sustainable development ethic, and Indian culture has helped make an activist environmental stance part of the tribe's Indian identity.

Perhaps the most startling message in this book is that sustainable development, at least on the Menominee Reservation, has little to do with an understanding of Marxist, capitalist, or any other kind of economics. Decision making about the forest is centered in Menominee Tribal Enterprise's Forestry Department, headed by Marshall Pecore, a lifelong resident of Menominee, who is also a trained forester. The Forestry Department is both a division of Menominee Tribal Enterprises, the business and sawmill management entity responsible for profiting from forest harvest operations, and the protector of the forest. As has happened so often in the past, decisions about what trees to cut and when to cut them are made by the rules spoken so eloquently when Charlie Frechette quoted an unnamed chief: "But take only the mature trees, the sick trees, and the trees that have fallen." The methods used to decide which tree to cut and which to leave alone in any one stand of trees are highly technical and based on a rigorous scientifically based understanding of the forest. The value of one type of lumber on the market versus the value of another type plays little role in the process even though Menominee Tribal Enterprises is a for-profit business, roundly criticized by tribal members when it fails to turn a profit. On the reservation, tribal members commonly admit, with both regret and chagrin, that the Enterprises is only partially a business. Its first duty is to the forest's welfare. Its second duty is to the tribal membership's welfare. Individual Menominee expect to see a monetary return from their ownership in Menominee Tribal Enterprises only when the first two duties have been addressed.

This does not mean that markets do not play a role in Menominee life. They play an enormous role. Menominee have the same appetite for consumer goods that other Americans have. A visit to any major store in Shawano, the town nearest to the reservation, on any Saturday night will reveal the number of Menominee taking full part in the American economy. The Enterprise is also always striving to make a profit through improving the efficiency of its operations, and it some-times succeeds.[1] But the forest and the attitudes and ideas constituting Menominee sustainable development (a long-term economic decision) are more important to the Menominee than any short-term economic decision. In the end, the forest is the people, and without the forest, as the reservation saying goes, the people will die.

2

What are the Elements of the Menominee Sustainable Development Model?

A Description of Menominee Sustainability

EIGHT COMPONENTS OF Menominee society have to be studied before the Menominee sustainable development model can be understood. These include historic, legal, cultural, spiritual, ethical, political, technological/scientific, and economic aspects of Menominee life. All of these aspects are interrelated in complex ways that often defy precise description. Overlaying these aspects are the management principles and practices that have made the Menominee Forest a sustainable forest environment.

These aspects of Menominee life exist as time-streams rather than as static elements. Discussions of sustainability must deal with time. The issues of intergenerational equity and preservation are integral to any successful definition of sustainability. Each of these eight aspects is in a constant state of change. Today's Menominee culture is not the Menominee culture of a hundred and fifty years ago. Therefore a description of any aspect of Menominee sustainability has to be offered with the caveat that not all or any one of them can be fully understood, measured, or described. One reason examples of sustainable communities are rare in the contemporary world is that any structure set up to maintain sustainability over long time periods must be flexible enough to accommodate changes that affect any society for a century and more.

Even more sobering than the realization that the task of exploring the Menominee model is difficult and complex is the fact that the

Menominee Indian Reservation in Northeastern Wisconsin is not a fully functioning sustainable society, at least not as that phrase was defined by the World Commission on Environment and Development. Sustainable development requires meeting major human needs and extending the opportunity to satisfy aspirations for a better life. Living standards beyond basic minimums are sustainable only if consumption standards take into account long-term sustainability (World Commission on Environment and Development 1987, 44).

The Menominee Reservation, in spite of recent economic gains, neither meets the major needs of all those living inside its expanding boundaries nor extends to all of its population "the opportunity to satisfy their aspirations for a better life." The reservation's problems with poverty, alcohol and drug abuse, social welfare dependency, lack of successful educational achievement by many students, child and spouse abuse, and crime, among others, are severe. Residents caught in the cycles of these social ills are not exceptions to a positive reltionship between the reservation's Indian population and its ability to provide for the needs of its people while, at the same time, providing opportunities for better lives. As impressive as Menominee sustainable development is, it has not solved the age-old problems of poverty.

The Menominee have also failed to achieve perfection as environmentalists. Beer can litter strewn beside a lake hidden deep in the Menominee Forest is not unusual. Menominee have also contaminated soils with leaking gas tanks, left chemical or non-biodegradable debris behind them after logging, and attempted to achieve personal profit while destroying part of what Menominee foresters are attempting to accomplish.

Still, even with these caveats, the Menominee sustainable development record is remarkable. The tribe has, over the course of more than one hundred years, been working out in the most practical, day-to-day sense possible, a set of sustainable development policies that have not only been pursued for a considerable period of time, but have achieved a marked degree of success.

Some elements of this success can be discerned by examining both the tribe's human and natural resources. If poverty and its attendant ills have not been banished from the reservation, the Menominee have still managed their natural resources and ecosystems in ways that have effectively balanced resource use and long-term resource sustainability. For a period extending through nearly eleven decades, the Menominee have made such wise use of their major resources that they exist not only within healthy, thriving ecosystems, but are also more economically productive today than they could have been eleven decades ago—even if the Menominee had possessed at that time the ability to fully use these resources.

In addition, even though full use of the Menominee resource base has not managed to support the Menominee people to the degree envisioned as sustainable by the Brundtland Commission, they have provided an economic and cultural base that has allowed the Menominee to aggressively search for solutions to their nagging, complex problems. These problems have not been easily solved. Lyndon Johnson's famous War on Poverty failed to reach its objectives on the rural Menominee Reservation in the same way it failed to alleviate poverty in America's inner cities. But the Menominee are aggressively facing up to their problems and trying to solve them. Signs of hope exist that one day the reservation will become a fully functioning sustainably developed society, meeting all the criteria for such a society outlined in *Our Common Future*.

This chapter does not construct a complete model of Menominee sustainable development. That is the task of the entire book. Rather, it starts the effort to (1) establish that the Menominee have created a sustainable development model, using the Brundtland Commission's definition of sustainable development, and (2) identify that model's elements.

Description of the Menominee Indian Reservation and Its Resources

The Menominee Reservation contains approximately 235,000 acres of land forty-five miles west, and a tick north, of the nearest city, Green Bay, Wisconsin. In the past the reservation's boundaries were conterminous with the boundaries of Menominee County, Wisconsin's smallest county, but today, because of recent tribal land purchases, the reservation bulges out of the county's borders into Shawano County, one of Wisconsin's most successful dairying regions. Menominee County itself is made up of ten townships: Township 28 North, Ranges 15 and 16 East; Township 29 North, Ranges 13, 14, 15, and 16 East; and Township 30 North, Ranges 13, 14, 15, and 16 East. The Menominee Indian people own most land within the borders of both the reservation and county, but a small, growing non-Indian community exists along the shores of man-made Legend Lake located near the reservation town of Keshena.

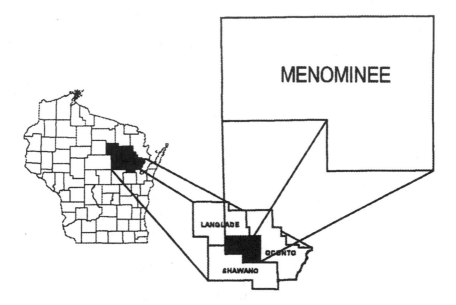

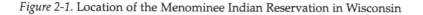

Figure 2-1. Location of the Menominee Indian Reservation in Wisconsin

Of the 235,000 acres of reservation land, approximately 220,000 acres are forested. The Menominee Forest is the primary natural resource owned by the Menominee. Major tree species include white pine, hemlock, sugar maple, red maple, red oak, basswood, and yellow birch, although an abundant mix of other species can be found. Most who visit the forest for the first time assume, unless they already know better, that it is pristine and untouched. Such is not even remotely the case. Vegetationally, the forest is in the Northern Hardwoods-Hemlock-White Pine Forest region of the United States.

The reservation's other major resource is water. An estimated 330 miles of trout streams; forty-four lakes ranging in size from Legend Lake, 1,304 acres, to Red Springs Lake, 1 acre; the Wolf River and its tributaries, the West Branch, the Little West Branch, the Evergreen, the Red; and the South Branch of the Oconto River grace the reservation. The Wolf River and its tributaries drain about four-fifths of Menominee County to the south. The South Branch drains the northeastern one-fifth of the forest to the east. Most lakes are small with only twelve lakes larger than fifty acres.

The Wolf River is one of the most scenic, dramatic rivers in the eastern half of the United States. Waterfalls and white rapids give way to stretches of calm, meandering movement as river waters flow past heavily forested banks marked by the dark and light of a variety of rock formations. Bald eagles, ospreys, a variety of duck species, swans and geese during the right season, herons, cranes, otters, beaver, crows, ravens, thrushes, chickadees, black bear, deer, and a host of other wildlife fly, swim, and frequent the riverine environment.

The rocky formations along the Wolf are formed of Precambrian crystalline granite and syenite of the Wolf River Batholith, which also underlies the forest's soils. This rock is a southern extension of the Canadian shield. Granite outcroppings can be found throughout the forest, especially along and within stream beds. In the reservation's extreme southeast, the bedrock is composed of quartz monzonite. The forest's soil is primarily glacial till and water-laid deposits of varying depth (Newman 1967; Soil Survey Division 1967). Hills undulate in a topography ranging in elevations from 800 to 1,425 feet.

In the northwest townships soils are primarily loam and silt loam soils. In eastern and southern townships soils change to sand loam and sandy. Sandy soils are primarily glacial outwash and are concentrated in the lake region. Peat swamps, both with and without forest cover, interlace the forest (Newman 1967).

Three historic communities are located on the reservation: Keshena, Neopit, and Zoar. A new community, Middle Village, on the reservation but outside Menominee County's borders, was established as a planned village in 1995. In addition, three areas have dispersed populations of different densities: Legend Lake, where the non-Indian population owns homes, South Branch, and West Branch. The 1990 census counted 3,411 people living within the reservation's borders. Keshena (population 685) is the largest of the three communities. Zoar, with a population of 115, is the smallest (U.S. Census 1990). There is evidence that the population is in the midst of a rapid growth spurt. In 1994 the tribe's enrollment office counted 3,539 tribal residents living on the reservation. This number would not include Legend Lake's non-Indian population (Menominee Indian Tribe of Wisconsin's Enrollment Office 1994).

The average general temperature is approximately 50 degrees Fahrenheit, with a range from 40 below zero to 105 degrees. The average frost-free season is only 126 days long. Annual precipitation ranges from 18 to 38 inches per year, with a 30-inch average. Dry seasons are common. Later in this book the role of dry seasons and how they have played a major role in the history of the development of Menominee as a sustainably developed society will be described. The average annual snowfall is approximately 47 inches. Snow covers the forest floor and the meadows during an average year for 110 days (Newman 1967; Soil Survey Division 1967).

The Menominee Landscape Compared with Surrounding Land

What distinguishes the reservation from the surrounding landscape is not its weather, rock formations, nor even its extensive water systems, but the 220,000 acres of the Menominee Forest. Just how unusual this forestland is can be seen by comparing the reservation to lands surrounding its borders. Three Wisconsin counties and one Indian reservation share its boundaries. Shawano County is to the south and west, Oconto County to the east, and Langlade County to the north. The Stockbridge-Munsee Reservation is no longer a block of land, but at least one parcel of Stockbridge-Munsee land abuts Menominee's western border. Also to the north is the Nicolet National Forest, which stretches over parts of seven northern Wisconsin counties.

All three counties conterminous to the reservation's borders have a mixture of farming, tourism, manufacturing, consumer, and forestry and forestry-related economies. The forests largely give way to interspersed woodlots as you travel south and farmland begins to dominate the landscape. North into the Nicolet National Forest forestlands dominate.

What makes the Menominee landscape remarkable is how plainly it stands out from surrounding land. Landstadt satellite photographs clearly reveal the Menominee Forest's boundaries on all four sides of the reservation. The forest density is even plain to the north where the Nicolet National Forest and the Menominee Forest meet. On one side of the line the Nicolet is sparser, with smaller trees. On the line's other side, Menominee trees rise above their second- and third-growth counterparts to the north, making an unmistakable boundary.

Scott Landis in an article entitled "Seventh-Generation Forestry" described the sensation of driving into the reservation from Shawano County for the first time:

> A web of blacktop stretches across central Wisconsin. From the Mississippi River in the west to Green Bay and Lake Michigan in the east, the roads are laid out in a surveyor's grid and named with letters of the alphabet. When the planners ran out of letters, they simply doubled them up. The towns and counties these roads stitch together bear the Indian and French names of an earlier era: Oshkosh

and Winnebago, Eau Claire and Marinette. East of Wausau, the nineteenth-century logging capital of central Wisconsin, county road N
slices a furrow through the countryside like the blade of a plow. A
short transit to the north on county road D, a jog to the east, and the
road leaves behind the tidy barns, painted silos and rolling fields of
Wisconsin dairy country. Descending a gentle grade, the two-lane
highway rounds a curve and is swallowed up in a wall of trees.
(Landis 1992, 27)

Newman, writing in 1967, is, in his technical way, as dramatic in
his description of the difference between the Menominee Forest and
the Nicolet National Forest as Landis is in describing the sensation of
driving from farmland into the forest. The first point Newman makes
is that the distinction between the two areas is not one created by any
natural differences. The differences "are predominantly due to cutting
practices and fire, rather than other environmental factors." He points
out the Nicolet was formed in the 1930s and has been under forest
management by the National Forest Service since that time. Then
Newman gets both technical and effusive:

According to Gary Koeppen, the district ranger (for Nicolet), the
allowable cut on the district is ten million board feet. Of this, approximately one million feet is sawtimber, and the other nine million is
pole-sized timber. The Menominee, with twice the land area, has an
allowable cut four times greater than the district, and a sawtimber
cut thirty times greater! The by-products cut is about equal.
    The Eagle River district (which contains the Nicolet) is fairly typ-
ical of the forest land in the northern Lake States which has been
receiving some management, and therefore, this comparison shows
that the Menominee Forest is an island of sawtimber in a region of
second growth hardwoods and aspen. (Newman 1967, 113–14)

Although today's Nicolet Forest covers more acres than at the time
of Newman's writing, and production numbers on both the
Menominee and Nicolet side of the forest line have changed, what
Newman recorded is still essentially the same. The Menominee Forest
is richer in larger trees, has a richer mix of species, and is more dense
than the Nicolet Forest. It also produces more board feet of timber per
acre for harvest as well as more board feet of commercial species.
    David J. Mladenoff and John Pastor, writing about the current sit-
uation in the Northern Hardwood and Conifer forest region, which
stretches from the northeast coast of both Canada and the United
States through far northern Minnesota, describe the Nicolet's general
situation after changes caused by the clear-cutting frenzy after the
Civil War:

The changes resulting from this forest removal dramatically altered regional ecosystems at all scales. Forest species composition and stand structure, ecosystem processes, and wildlife habitat were all changed (Flader 1983). . . . The net result of logging is a homogenization of stand ages and disturbance intervals across the boreal-northern hardwood/conifer regions.

Also during this century, populations of forest animals that have key roles in natural ecosystems have been severely reduced or eliminated, either through habitat change or directly through harvesting.

Successional forest dramatically increased habitat for white-tailed deer . . . while human hunting and trapping pressure eliminated or drastically reduced moose . . . other large ungulates, beaver, and carnivores, including the timber wolf. (Mladenoff and Pastor 1993, 150–54)

This dramatic simplification of the Nicolet's ecosystems compared to Menominee ecosystems has left even-aged stands of trees with fewer species. Even wildlife diversity has lessened.

This is true even though the forestry and logging practices in the Nicolet are today not that different from Menominee Forest practices. Clear-cutting is not the norm, and the U.S. Forest Service is certainly aware of good silviculture and works hard to maintain the Nicolet's overall ecosystem health. Still, because of the political realities that face any U.S. government agency, the Forest Service is much more conscious of multiple-use policies than the Menominee, who base nearly all their decisions on their understanding of sound silviculture practices. This has made management of the Nicolet more complex than Menominee Forest management. The effort to please a broader public has also led to decisions more firmly based in economics than in good silviculture. Even though many silvicultural and harvesting practices in the two forests are the same today, a large difference exists overall.

In a report for the U.S. Department of Agriculture on biodiversity risks facing both the Nicolet and Chequamegon National Forests in Northern Wisconsin, Thomas R. Crow, Alan Haney, and Donald M. Waller cataloged the risks facing these forests. Included were:

habitat fragmentation; disruption of ecological processes (especially historical disturbance regimens); modifications of microsites needed for regeneration and establishment of many species; pervasive modification of forest structure and composition; and genetic shifts in populations, ranging from the introduction of exotic species or alleles to the reduction or loss of native species or alleles. Fragmentation results from many causes, including the presence of roads and utility rights-of-way, the construction of dams and housing developments, the growth of trees in originally open areas, and the dispersion of timber harvests big enough to cause openings through once continuous forests. These activities also boost the amount of edge habitat, precipitating edge effects that separately threaten some elements of diversity. Populations isolated by habitat loss or fragmentation face further demographic and genetic risks, including reduced opportunities for local demographic rescue, decreased population movement along environmental gradients, and increased inbreeding and drift. (Crow, Haney, and Waller 1994, 1)

This passage outlines not only the problems and risks present in the two national forests, but also describes how the problems created by the boundaries between the Menominee Forest and dairy farms, and even the boundary between the Nicolet and Menominee Forest, impacts the Menominee Forest. Ecological systems do not exist in isolation from lands connecting with them. Ecosystems on contiguous lands affect the bordering ecosystem. The clearest example of this can be found in examining large mammal populations. The reservation has a healthy black bear, white-tail deer, and beaver population, but moose or wolves are rare to nonexistent.[1] Wolves have been successfully reintroduced into Northern Wisconsin. Some packs have entered the Nicolet after having been extinct in the region for many decades. At one time the Menominee Forest contained a larger diversity of large mammals than at present. Large mammals such as moose and wolves need larger ranges than the reservation can provide.

In their discussion on diversity, Crow, Haney, and Waller also point out that:

Diversity may also be threatened by changes in forest composition and structure that affect ecological processes. Northern Wisconsin forests today are younger, more even-aged, and contain more early successional trees, fewer tip-up mounds, and less coarse wood debris than the mostly old-growth forests they replaced. Forests have been further modified by favoring tree species with rapid growth and high economic value and by suppressing fires that were once widespread. Because the maintenance and regeneration of many species depend on these aspects of forest structure and historical disturbance regimens, these changes have brought fewer niche and microsite opportunities for those species dependent on old or dead trees, exposed mineral soil, or conditions associated with fire. (Crow, Haney, and Waller 1994, 1)

As will be shown later, the Menominee Forest has fewer of these problems than either the Nicolet or Chequamegon. Historical regimens did occur that modified parts of the Menominee Forest by "favoring tree species with rapid growth and high economic value."

Fires have also been suppressed, causing both maintenance and regeneration problems with species like the white pine.[2] Insect infestations, disease, and other problems that start in either the Nicolet or Menominee Forest or along their borders easily spread into the other forest. Species mix on the two forests' border is impacted by the mix just across the line. The same statements can also be made about the many woodlots on dairy farms that abut the Menominee Forest. Forests, woodlots, and farmland constitute living systems that affect each other both negatively and positively. Nonpoint pollution originating from area farms is not a respector of deed boundaries.

Still, the general rule stated previously holds. The Menominee Forest is richer in the multiple-age tree stands, niche areas such as tip-up mounds and other vegetation and is more diverse than the Nicolet Forest contiguous to it. Compared to surrounding lands, the Menominee Forest is so distinctive that its outlines on Landstadt photographs from space are as plainly marked and regular as lines drawn on a map.

The Significance of What the Menominee Have Achieved

The best way to establish the significance of what the Menominee foresters have achieved is to study an unpublished doctoral thesis written by James Gilbert Newman. In his thesis Newman studied the volume distribution of saw-timber species occurring on the reservation as they existed in 1914, when the first cruise survey was completed, and in 1963 after several decades of sustainable forestry practices. "The most impressive gain" from 1914 to 1963, according to Newman, "is the gain in volume" in the forest.

> The 1963 sawtimber volume shows an increase of 70 percent over that in the 1914 survey. The pole-sized timber gained even more in this period, with the 1963 cordwood volume being almost 9 times as large as the 1914 volume. If the cord volume is converted to its equivalent in board feet (2 cords = 1,000 board feet), and added to the sawtimber volume, the net increase over the 50 years has been approximately 1,502,761 M board feet, or a net per acre annual increase of approximately 160 board feet. During this same period approximately 971,188 M board feet were harvested, for an average of 90 board feet per acre over the entire forest. Thus, net increase plus cut equals 250 board feet per acre per year, for an annual production of approximately 56,000 M board feet. (Newman 1967, 111 and 113)

Table 2-1
Forest Net Growth
Sawtimber, Soft Woods

Species	1914		1963	
	Million Board Feet	Percent of Volume	Million Board Feet	Percent of Volume
White Pine	155,626	17.3	324,076	21.2
Red Pine	20,131	2.2	30,528	2.0
Jack Pine	102	—	3,621	0.2
Hemlock	491,538	54.7	367,093	24.1
Cedar	3,630	0.4	30,971	2.0
Spruce (Black & White)	350	—	4,101	0.2
Balsam fir	131	—	788	0.1
Tamarack	4,904	0.6	860	0.1
TOTAL	676,412	75.2	762,848	49.9

Sawtimber, Hard Woods

Species	1914		1963	
	Million Board Feet	Percent of Volume	Million Board Feet	Percent of Volume
Sugar Maple	62,597	7.0	263,753	17.3
Red Maple	0	0	16,324	1.1
Yellow Birch	60,738	6.8	105,623	6.9
Basswood	54,641	6.1	93,745	6.1
Elm (all)	28,998	3.2	115,780	7.6
Red Oak	8,006	0.9	69,945	4.6
White Oak	0	0	14,890	1.0
Beech	4,046	0.5	27,219	1.8
Ash (all)	2,590	0.3	14,122	0.9
Aspen	58	—	34,050	2.2
Miscellaneous	75	—	9,281	0.6
TOTAL	221,749	24.8	764,732	50.1
TOTAL Soft/Hard Woods	898,161	100.0	1,527,580	100.0

Table 2-2
Pole-Sized Timber

Type Group	1914		1963	
	Cords	Percent	Cords	Percent
Hard Woods	208,152	92.0	1,329,617	67.4
Soft Woods	18,165	8.0	643,383	32.6
TOTAL	226,317	100.0	1,973,000	100.0

Sources:

1914 data - Menominee file, Federal Records Center, Chicago, Ill

1963 data - Menominee Enterprises, Inc., Neopit, Wisc. (Newman 1967, 112)

"M" in this instance stands for million. What this indicates, of course, as is shown by the preceeding tables, is that the forest's net growth increased from 1914 to 1963 even though it had been intensively logged during that period using sustained yield,[3] selective cutting techniques.

The two trends these tables illustrate, the increase in the forest's sawtimber productivity and the change in species mix and volume, have continued to this day. The overall tree volume has also remained constant, if not improved. As Marshall Pecore pointed out in a 1992 article:

> When the Menominee Reservation was established in 1854, there were an estimated 1.5 billion board feet of sawtimber growing stock. From 1865 (when annual cut volumes were first recorded) nearly 2 billion board feet of sawtimber had been harvested. The most recent inventory indicates that sawtimber stocking is still at least 1.5 billion board feet, even after 138 years of harvesting this same acreage. (Pecore 1992, 16)

Therefore, the Menominee have managed to sustainably manage their forest for 141 years to this point. At the same time they have fully utilized the forest, increasing its commercial productivity while maintaining its overall health as indicated by the abundance of two Northern Hardwoods-Hemlock-White Pine Forest region indicator species, hemlock and Canadian yew (Huff and Pecore 1995, 19).

This record can be compared to that of the Nicolet. The Nicolet encompasses 973,000 acres, of which 655,000 acres are categorized as national forest lands. From a forest three times the size of the Menominee Forest, only 35,466 acres were harvested in 1983. Allowable sale quantities from this acreage included 26.6 M b.f. (million board feet) of sawtimber and 70.4 M b.f. of other products. This compares to the available softwood and hardwood on Menominee of 1,527,580 M b.f. Admittedly, Nicolet's multiple use policies do not allow as intensive a cut as the Menominee practice, but the volume available for harvest in the larger forest is also much lower.

Management of the reservation's water and land resources has been, with the exception of the construction of the Legend Lake Dam, equally as impressive. In addition to water use for home and community, the reservation's water systems have had several different historical uses. These have included transportation for both people and logs, recreational uses, fishing uses, and power generation. Today primary uses center around water consumption, fishing, and rafting through the Wolf's scenery and whitewater.

According to a recent U.S. Geographical Survey study, "there is no widespread problem with respect to high concentrations of health-related inorganic constituents in groundwater on the Reservation." The study found that "the principal ground-water-quality problems on the reservation that affects use of ground water for domestic water supplies are iron and manganese concentrations . . . " a common problem in north-central acquifers. The only problems associated with these particular, naturally occurring problems are "objectionable tastes, staining of laundry and plumbing fixtures, and clogging of well screens and distribution systems." (United States Geological Survey 1994, 10).

The U.S. Geological Survey also found that:

> Water samples collected from Reservation streams and lakes were similar to ground water (calcium magnesium bicarbonate type), reflecting the strong surface-water/ground-water interaction in the study area. The chemical composition of water from lakes having inlets or outlets indicate that they are less affected by precipitation than are lakes that do not have inlets or outlets. (United States Geological Survey 1994, 12)

In spite of heavy uses, including the influx of thousands of rafters each season on the Wolf River, the reservation's waters have much the same quality they had over a hundred years ago. Any visit to Big Eddey Falls on the Wolf, with its roar of whitewater through narrow granite into the pool beneath the falls, both sides of the river lined by white pine and a healthy mix of both softwood and hardwood trees, is as convincing of this fact as any set of tables or statistics. The reservation's riparian ecosystems appear to be in excellent shape. Although extensive studies have not been completed, the mix of animal life, tree species, and plant communities is rich and diverse. If you are a rafter feeling the powerful sweep beneath you as water sprays your face and you cling to your raft as you half-fall, half-rush into the pool below Big Eddey Falls, you quickly become convinced of the purity of the landscape and water.

What makes this purity significant is that the landscape has been used for production since the first timber harvest by the Menominee in 1865.[4] The Menominee, from the time they settled permanently on the reservation in 1854 until the present, have avoided the tragedy of the commons described by Garrett Hardin (1980). They have held land in common throughout this period, but they have not used up its resources. They have, instead, increased the value of their resources while deriving common economic benefits from their produce. This is a record rare in today's world. More often than not, exactly the opposite occurs, and commons resources follow the path outlined by

Hardin. One resource consumer increases his or her share in order to derive a larger personal profit. This leads to a frenzy of resource use with each user increasing his or her take until the resource is destroyed.

The Historical Devastation of the Great Lakes Region Forests

A description of how the Great Lakes Region Forests were destroyed helps to put the Menominee Forest in context. It also helps to establish the significance of the Menominee achievement. Once, the lands surrounding the reservation were covered by forests as rich and diverse as today's Menominee Forest. These massive climax forests covered millions of acres, broken only by occasional clearings. Ancient white pines towered above river and stream banks, maple, oak, birch, hemlock, elm, basswood, black walnut, jack pine, tamarack, spruce, cedar, and other species and sub-species covered the hills, swales, and level land. The forest's heart was perpetual twilight, sun filtering through webs of branches and leaves. The forest understory was rich with life.

William G. Robbins said in his classic U. S. lumber industry study that

> every eyewitness account tells us that these timber stands were impressive, perhaps beyond imagination. One Forest Service study estimates that the original commercial forest in what is now the United States may have contained 5.2 billion board feet. A geographer writing in the 1920s believed that the virgin forest of the United States covered 821,800,000 acres. . . . Forests, in fact, were the predominant feature of colonial America. (Robbins 1982, 3)

What happened to this massive ecosystem stands in stark contrast to the Menominee ethic developed during the period when the forests surrounding the reservation were destroyed. Robbins described the destruction:

> In the last decades of the nineteenth century resource entrepreneurs, railroad promoters, lumbermen, mine owners, land speculators, cattlemen, and farmers indulged in the uninhibited and feverish exploitation of America's natural resources. Urged on by a ruthless and excessively competitive economic environment, these groups despoiled rivers and streams, devastated timberlands, wasted mineral resources, and overgrazed the western range. Robert Wiebe calls it "a rape of Gargantuan proportions." This wasteful pattern of land and resource use was part of the post-Civil War economy in which

> entrepreneurs took advantage of rising land values to liquidate and
> capitalize on the nation's virgin treasures. (Robbins 1982, 16)

The result was a changed national and Great Lakes regional landscape. The richness of the forests gave way to a landscape designed to
further the entrepreneurial, industrial, and commercial economies
present today in counties surrounding the reservation. The contrast
between reservation and non-reservation landscapes mirrors the contrast in ethics, culture, religion, and economic sensibilities that has
existed between the Menominee and their neighbors since the first
treaty between the United States and the Menominee Nation was
signed in 1817.

The Best Demonstration of the Menominee
Environment's Value

Perhaps the best demonstration of the Menominee environment's
value lies in an examination of the process that led up to and then initiated the disastrous termination experiment long after the Great
Lakes forests had given way to dairy farms, small towns, and paper
mills in the Menominee/Shawano area. During the planning phases of
termination, before the 1961 implementation date when the tribe's
treaty rights were abrogated, Congress agreed to place extraordinarily
strict limitations on the sustained yield cut of the forest. The State of
Wisconsin insisted the forest was an aesthetic Wisconsin, as well as a
Menominee, asset and should therefore be protected from potentially
negative actions by the Menominee in their effort to become assimilated into the American economy. This restrictiveness later led to a
successful law suit, *Menominee Tribe of Indians v. The United States of
America*, settled in 1984. Early in the termination process Wisconsin
politicians suggested that the reservation should be made a state park.
During the 1956 gubernatorial campaign the successful candidate,
Vernon W. Thompson, caused concern on the reservation when he
announced he believed the state should purchase the reservation for a
park. By 1956 the reservation represented a priceless historical artifact
that showed how the entire State of Wisconsin had once appeared. Its
forests were the largest northern hardwoods-hemlock-white pine forest type left in the United States, and its white pine stands were especially impressive because of the near impossibility of achieving white
pine regeneration after the virgin white pine forests had been cut.
    What led to this value for both Wisconsin and the Menominee was

the Menominee practice of sustainability. While "the nineteenth century resource entrepreneurs, railroad promoters, lumbermen, mine owners, land speculators, cattlemen, and farmers indulged in the uninhibited and feverish exploitation of America's natural resources" (Robbins 1992, 16), the Menominee developed the science and technology that has allowed them to sustain their forest environment without serious degradation for 141 years.

The question arising from this discussion is: What has led the Menominee to sustain their environment? The answer has already been partially given. This chapter begins the exploration of the eight aspects of Menominee sustainability that, joined together, have helped the Menominee achieve a sustainable community based upon the twin principles of sustainable resources and economic development (or economic well-being). Before a full understanding of the Menominee phenomenon can be achieved, however, each aspect must be explored in depth, and the contribution of each to the whole must be examined.

3

The Beginnings of Sustainability

An Aspect of Menominee History

THE FIRST STEP in understanding what has led to Menominee sustainable development is to examine Menominee history and then identify the historical events that have played a major role in shaping Menominee thought and behavior. This is not an easy enterprise since Menominee history is as long and complex as the history of any people who have survived for over five thousand years. Still, Menominee history gives, at the very least, some clues about why the Menominee developed sustainable development practices.

This chapter examines early Menominee history and identifies those events and historical movements that played a role in leading the tribe toward sustainable development and away from the American frenzy of entrepreneurial resource use in the 1800s. Later chapters explore how different aspects of Menominee sustainable development arose from the Tribe's history.

The reason a historical approach has been taken in this section of the book to explore what is, at heart, a contemporary policy issue is that the sustainability idea is anchored in time. The first test of any sustainable development policy is, Can it last? Is it sustainable given the changeability of human nature, human affairs, and the relationship of that nature and those affairs to the earth and its resources? If the policy cannot last for prolonged periods of time, then it is not sustainable.

Historical analysis, then, has an important role to play in understanding a society and culture that has the ability to work on both sustainable and development issues without allowing development policy to destroy sustainable policies. Historical analysis of the Menominee proves that sustainable policies have been in place for a period of time sufficient to truly describe them as sustainable. History, at least as it is used in this book, is also a source of data about the antecedents of sustainability as the Menominee have chosen to implement it, the general elements of sustainable and development policies, and the broader question of markers. How can policies that might lead to sustainable development in the larger society be identified as sustainable development policies? How can these policies be differentiated from policies that have a short shelf life even though they are labeled sustainable development policies? Do workable sustainable development policies have characteristics that "mark" them and make them identifiable?

This chapter begins with a description of the Menominee people. No history can be understood without an acquaintance with the people the history describes. Then it goes on to explore the history that led to early Menominee decisions to protect their forest resource forever rather than to use the resource to help ensure individual futures for the Menominee then alive. The major lesson to be derived from this chapter is one of motivation. In the late 1800s and early 1900s the Menominee had reasons for trying to preserve what they could from the disasters that had befallen them. The idea of sustaining the Menominee Forest while using its productive capabilities to economically develop the tribe did not occur by accident. It arose from a series of historical events that had left the tribe devastated and demoralized. To the Menominee of the time the idea of sustainability gave them a reason to hope for a better life.

The Menominee People

The Menominee's current reservation is close to where French explorer Jean Nicolet found the Menominee living in 1634 when he first canoed into the bay of Lake Michigan now known as Green Bay. Located only forty-five miles or so northeast of the modern Wisconsin city of Green Bay, the Menominee can fairly claim to still be living in the homeland they occupied before the coming of the French and other Europeans.

The Menominee are an Algonquian people most closely related to the Anishinaabe (Ojibwa), although in pre-European-contact times they had considerably more commerce with the Winnebago than with the Anishinaabe. The Menominee name comes from the Menominee word for wild rice, a dark grain that grows wild throughout Northern Wisconsin, Michigan, and Minnesota. Tribal legend has it that the Menominee earned their name because wherever they went wild rice started to grow in abundance, and when they stopped living in a place, the wild rice harvest dwindled.

Evidence indicates the Menominee observed an equality between the sexes before the more patrilineal society, in place by the time the Europeans discovered the Menominee, developed. Walter J. Hoffman, an ethnologist working for the Bureau of Ethnology and the first to write a book about the Menominee in 1896, noted that "some of the ancient customs respecting the disposition of property and children, in the event of the death of either parent, are still spoken of, though seldom, if ever practiced." One of the most important of these ancient customs was that "descent was in the mother's line, at her death both children and personal effects were transmitted to the nearest of the mother's kin" (Hoffman 1896, 43). Beaver Woman, after the bear and the eagle, was also the third Menominee totem created by the Great Spirit and, as such, was considered one of the tribe's most important totems.

Primarily hunters and gatherers before the Europeans came, the Menominee proved their adaptability with the coming of the French fur trade, then the ascension of the British to predominance over the Great Lakes region, and then the American ascension. During this period they normally were able to ally themselves with the most powerful force then contesting the region, sometimes against the efforts of other Indian tribes to stop the white man's advance. As a result the Menominee often managed to prosper better than other North American tribes. They also managed to maintain their Wisconsin residence, no small task in the face of changing economic conditions and the insatiable European, then American, desire for land.

Most accounts of the Menominee during their early relationships with the Europeans described them as industrious, well-made, fair, handsome, and with a mild, but independent, demeanor (Hoffman 1896, 34). Alfred Cope, a Quaker emissary from President Andrew Jackson to the Menominee, found these descriptions accurate even after the Americans had gained predominance over Wisconsin. The Menominee lands were being "sold" to the United States government as the result of a series of treaties forced upon the tribe. Cope and Thomas Wister, the commissioner sent to the Menominee by Jackson, expected a drunken, disorderly, hard-to-manage people that would give the commissioner and Cope endless trouble as they worked to implement the 1848 treaty signed at Lake Poygan. Cope recorded in his diary what he observed instead:

> One thing was remarkable among them, which could hardly have been assumed for the occasion—the universal decorum which prevailed. It gave evidence of good discipline or an intuitive attachment to order and harmony. No squabbling children, contentious youth, or scolding wives were to be found in the company. . . Strict honesty prevailed with all ages. Poultry, various implements and household utensils, very desirable to these needy people, were exposed night and day, yet not one of them was taken or even displaced, so far as was observed. The lodgings of the Friends were not locked, the front door being kept shut at night by a stick, the lower end of which rested against a small projection from the entry partition. No attempt at intrusion was ever made. Could one have lived in a white community of the same number for the same time, selected at random from the destitute, and been able to render as good an account of them? (Cope 1967, 127)

Cope also describes the hospitality and generosity of the Menominee with whom he and Wister completed their business in Green Bay (Cope 1967, 127).

As was indicated in chapter 2, today's Menominee are much more troubled than their ancestors. Poverty rates are high, drug and alcohol abuse is a constant presence, and a variety of crime, including theft and murder, plagues the people of the wild rice. Still, traditional Menominee heritage runs deep, and a constant effort is made to improve tribal members' lives through a variety of development efforts. Most tribal members possess their ancestors' integrity. During recent years dramatic improvements have also been made in many aspects of Menominee life.

This chapter's purpose, however, is not to explain the Menominee people and how they have arrived into their current social condition. Rather, it identifies the historical events and movements that have led to the development of the tribe's sustainable practices.

The Importance of Land to the Menominee

To contemporary Menominee, one of the most important aspects of sustainable development is centered in the importance the tribe puts on the tribe's land base. The history of why land has become so important to the Menominee explains why the Menominee place such a high value on their sustainable development policies. Land, as will be shown, is not an economic unit that can be bought and sold in the free marketplace. Rather, it is communally held, not only for this generation of Menominee, but for all the generations that have lived and all of those yet to be born.

The important part of Menominee history and the reservation's land starts with the American government's ascendancy over Menominee territory after the War of 1812. The Menominee had, in many ways, dealt successfully with both the French and the British from the time of Nicolet's Green Bay shore landing. From the early 1600s through the early 1800s the Menominee lived on ancestral lands, hunting, fishing, gathering, and trapping, modifying their society only to suit their interest as they became familiar with the ways, tools, and economy of Europe.

The Americans were not like the French and English, however. The French and English had used military force against Indian people when it was to their benefit, but they had also, to a large measure, been willing to let the Northern Woodland tribes live wherever and however they wanted to live. They built forts and small villages but did not come to the region to populate it. As long as the French and British could derive profit from the Great Lakes tribes, they were willing to allow the tribes their homelands.

What made the Americans different was that they came to Menominee territory following the manifest destiny doctrine. This doctrine claimed America's destiny was to dominate the North American continent from sea to shining sea. The Menominee lived inside the American continent's borders. They therefore had to be brought under government control. Menominee land was suitable for the production of timber and practice of agriculture, and the Americans had a God-given right to any resources they could make their own.

This particular approach to what became known as "the Indian problem" ushered in the treaty era of Indian/American relations. In 1815 the Menominee dominated a large part of present-day Wisconsin. According to a United States government official who visited Green Bay in 1820,

> The Menominees claim the whole of the waters of Green Bay, with its islands. On its north-west shores, and on Fox river, they claim from the entrance of Menominee river, in length, one hundred and twenty

miles, south-west and north-east; and in breadth sixty miles. On the
south-east shore of the Bay, and on Fox river, from the river Rouge,
on Red River, to the Grand Cockalaw, a distance of forty-five miles,
and twenty-four in breadth. (Hoffman 1896, 21)

This territory would have encompassed over eight thousand square
miles of mostly heavily forested land, a magnificent wilderness.

The Menominee did not, of course, populate this wilderness with
small villages and homesteads. They hunted, fished, and gathered
throughout their territory during different seasons at different times.
The way they survived, in a cold climate, made such a large territory
necessary. If the wild rice harvest where they chose to locate turned
out to be poor, or if the sturgeon did not spawn on the Wolf River in
normal numbers, or if hunting was suddenly poor on the Sturgeon
Bay Peninsula, or if a hostile band of another tribe was passing
through and dictated a wide berth, then the tribe's band harvesting
rice, spearing sturgeon, hunting game, or avoiding an enemy band
would move to another place for a few weeks, a season, or even longer
time periods.

When the treaty period began in 1827, it not only marked the
beginning of the loss of most Menominee territory, but it also ended a
way of life. The new life imposed upon the tribe by the Americans led,
eventually, to hard times. At times during the process of losing land
and coming to live on the present day reservation, the Menominee
believed their days as a people were ending in disease and starvation.
Poverty's degradation and the loss of hope became interlinked with
who the Menominee were as a people. Inevitably the treaties not only
took away Menominee territory, but were also a part of the process that
first threatened the tribe's continued existence and then impoverished
the tribal members who survived. In time the loss of land came to be
integral to the sense that the Menominee had lost their "Indianness,"
the core identity that made them who they believed they were.

The treaty period started inauspiciously enough, at least from the
Menominee viewpoint. From 1818 to 1820 Colonel John Bowyer, the
Indian agent at Green Bay, negotiated a treaty requiring the tribe to
cede more than a million acres for a total of eight hundred dollars, or
a penny for every twelve and a half acres. Bowyer did not negotiate
with the tribe's leaders. This resulted in a Menominee victory. When
Menominee leaders protested vigorously to the government, the
treaty was not submitted by the president to the United States Senate
for ratification. It was later canceled by President James Monroe.

Throughout the treaty period the Menominee continued to use similar protests as a way to escape the inevitable. The most significant treaty signed by the Menominee, the Treaty of the Wolf River (1854), restored acreage that makes up today's reservation after the tribe had lost all their ancestral lands as the result of the 1848 Treaty of Lake Poygan. This achievement resulted from the same pattern of complaints and persistence that characterized the Menominee reaction to the treaty negotiated by Bowyer.

Complaints and persistence were not always successful as tactics. From 1827 through 1856 the Menominee land base was steadily eroded. Cession treaties were signed in 1827, 1831, 1836, 1848, and 1856. Living in 8,000 square miles of wilderness, the Menominee were forced to accept a land base of 235,000 acres. Independent people who had mastered life in a harsh environment became wards of the United States government. They scored significant victories along this path. The Treaty of the Wolf River was significant in that it allowed them to continue in their homeland. Not many Indian tribes achieved this victory during those years. They also won concessions in several treaties that would later strengthen the tribe's many lawsuits against the government. But most of these victories were, at the time, bitter, little victories that did not lead toward a shining tribal future.

Menominee bitterness can best be explained by describing events surrounding the 1848 Lake Poygan Treaty. The negotiations occurred right after the Black Hawk War in southern Wisconsin where the Menominee had fought with U.S. troops. As a Captain Marryat, present at the negotiations as part of the U.S. government team, was led to remark:

> The Indians . . . are compelled to sell—the purchase money being a mere subterfuge, by which it may appear as if the lands were not being wrested from them, although, in fact, it is. (Keesing 1987, 140; *Wisconsin Historical Collections* xiv, 139)

The government's clear intent was to remove the Menominee from Wisconsin to what is now Minnesota in order to make room for the rising tide of non-Indian settlers. The Menominee objected with bitterness. A French settler in Lake Poygan related the following conversation overheard between a Menominee chief and other Menominee:

> You don't expect he has come to decorate your ears with silver ear bobs? No, he comes simply to get the balance of our country! . . . he

proposed to move us across the Mississippi . . . he says there is an abundance of all kinds of game there; that the lakes and rivers are full of fish and wild rice.

(His listeners interrupt: "Why doesn't he go himself and live in such a fine country!")

You know how the Kechemocoman (Great Knife, as they named the American) never gets rebuked at a refusal; but will persist, and try over and over again till he accomplishes his purpose. (Keesing 1987, 228)

The Menominee negotiated for days, causing one U.S. official to storm away from Lake Poygan, "angry as well could be, having lost all hopes the treaty could be made" (Keesing 1987, 140–41). But in the end the treaty was ratified in January of 1849. The Menominee ceded, sold, and relinquished forever all their lands remaining in Wisconsin in return for not less than six hundred thousand acres in what is today tourist resort country in Minnesota. The Menominee signed the treaty only after they were threatened with forcible removal.

Still, they managed an important victory. Before they put their marks on the document, they insisted they should have the right to tour the Minnesota lands at the U.S. government's expense. If they objected to the place Kechemocoman (the Great Knive) wanted to send them, then the treaty was rejected (Keesing 1987, 141). This provision, along with the bitterness of their complaints and persistence, allowed them their Treaty of the Wolf River victory.

The Loss of Well-Being, the Loss of Forest

The result of Menominee land loss was tragic. In return for "selling" their lands in the 1848 treaty, the Menominee were to receive provisions to help sustain them while they settled their new reservation. That portion of the treaty was implemented after the Menominee settled their present reservation. George Washington Ewing and his partner, Colonel Thompson, received the contract let by the U.S. government for providing provisions. These provisions included bad food and rotten meat. Most of the provisions never arrived on the reservation. Beck explains the situation as well as any of the Menominee historians:

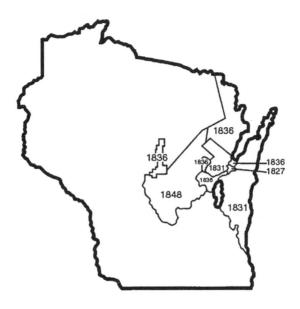

Figure 3-1. Map Showing the Dates When Treaties Forced Menominee Land Cessions. (Adapted from Lurie 1980, p. 15.)

Table 3-1

Menominee Treaties and the Year They Were Signed

Treaty of St. Louis	1817
Treaty of Prairie des Chiens	1825
Treaty of Buttes des Morts	1827
Treaty of Washington	1831
Treaty of Cedar Point	1836
Treaty of Lake Poygan	1848
Treaty of the Wolf River	1854
Cession to the Stockbridge-Munsee Tribe	1856

Many of the federally furnished provisions had spoiled before they arrived. That winter many Menominees died of malnutrition; it was the first of several winters of starvation which, combined with the ravages of disease, reduced the Menominee population from 2002 at the time of removal to 1336 by 1870. (Beck 1994, 441–42)

As late as 1820 Reverend Jedidiah Morse, as part of an Indian nations tour taken at the request of Secretary of War John C. Calhoun, estimated the tribe had a population of 3,900. In 1834 one quarter of the tribal membership died from smallpox.

At the same time the Menominee were attempting to understand and deal with these losses, another kind of loss was occurring all around them. As they gave up one large chunk of their territory after another, they saw the disappearance of the forests they had lived in for all the tribe's generations. The great pine baron companies moved from land parcel to parcel. Farms and crops grew behind the saw's blade as quickly as American entrepreneurism and greed could make them grow. The great pre-American-settlement Great Lake forests with their diverse tree species and complex community patterns fell to ax and saw. Heterogeneity, augmented by natural disturbance, with stands varying in "age from patches of young trees developing after a blowdown and even-aged communities originating from fire, to mature stands of pine, hemlock, and various hardwoods, with individual stems ranging in age from 250 to over 400 years" (Stearns 1987, 27) disappeared.

The Menominee saw the forest's disappearance in a different way than most Americans, who celebrated the wealth created by the wholesale harvest of the continent's natural resources. Louise Erdrich, a contemporary Anishinaabe fiction writer, describes the woodland people's sense of unreality as their land was changed in ways they had never imagined:

It began as a far-off murmur, a disturbance in the wind. We noticed an unusual number of birds and other animals that nested or burrowed in trees. Thrashers and grouse settled in the wild grass around Fleur's cabin. Kokoko silently appeared in broad daylight and walked the roof at dusk, uttering one note. Rabbits came to the edge of the clearing, squirrels bounded through the leaves, fighting pitched battles over territory. The murmur grew more distinct.

Then one day we could hear them clearly. Ringing over the water and our shore came the shouts of men, faint thump of steel axes. Their saws were rasping whispers, the turn of wooden wheels on ungreased axles was shrill as far-off flocks of gulls. (Erdrich 1988, 206)

Rasping whispers that mowed down forests and left farms and cleared land in their wake brought poverty and dependency. The mark left on the Menominee and other tribes was deep and painful. A holy Wintu tribal woman from California spoke in the late 1800s as eloquently as any woodland Indian about the destruction of the plentiful forests:

> The white people never cared for land or deer or bear. When we Indians kill meat, we eat it all up. When we dig roots we make little holes. When we build houses we make little holes. When we burn grass for grasshoppers, we don't ruin things. We shake down acorns and pinenuts. We don't chop down the trees, kill everything. The tree says, "Don't. I am sore. Don't hurt me." But they chop it down and cut it up. The spirit of the land hates them. They blast out trees and stir it to its depths. They saw the trees. That hurts them. The Indians never hurt anything, but the White people destroy all. . . . How can the spirit of the earth like the White man?. . . Everywhere the White man has touched it, it is sore. (McLuhan 1971, 15)

Most Americans saw the great Upper Great Lakes forest destruction as progress and a part of the great entrepreneurial explosion of enterprise so much a part of 1800s American life. The Menominee and other tribes viewed it as disaster. They lost their land, independence, and well-being as Americans forced their will upon Indian people. As the forests disappeared, the chance for returning to the life the Menominee had fashioned for themselves over centuries of time disappeared also. The Menominee now had to live within reservation boundaries. They had to wrest a living from a land minuscule in comparison to the lands that had once provided sustenance. The loss was overwhelming and bitter. On the reservation, voices wondered if the tribe could survive the catastrophes that had befallen the people of the wild rice.

By the late 1800s four major symbols were interlinked in Menominee thought: the loss of aboriginal land, the loss of forest, the loss of independence, and the growth of dependence on the American government, and poverty. Disease, starvation, and brutal living conditions had, in one generation, become the norm rather than the exception. The great question of that time, at least to the Menominee, was, What kind of future faced the tribe?

Farming and the Effort to Become Loggers

The United States had a clear answer to this question. The future of the tribe and its families was the American future. The Menominee would become farmers and not only feed themselves, but help to feed the growing nation. They would turn from the pursuits of savages and become civilized in the ways of European civilization.

This effort began in earnest with the 1831 treaty. In the treaty the Menominee granted half a million acres to members of several eastern tribes including the Oneida, the Stockbridge, the Munsee, and the Brotherton. They ceded lands on Green Bay's eastern shore south to the Milwaukee River and east of Lake Winnebago to the government. In return the tribe received annuity payments, farming equipment, schools and a teacher for educating them in European ways, and the labor of farmers, blacksmiths, and millers. The laborers were supposed to help the Menominee learn these valued trades so they could survive in a civilized manner. The tribe was also given the services of an Indian agent who was to act as an intermediary between the tribe and the federal government and help pave the way for integration into American life.

This early experiment failed for a number of reasons. Corrupt Indian agents, as well as corruption among those hired to teach the Menominee to farm, reduced Menominee assets. Americans wanted the land the Menominee had chosen to settle on when the farms, mill, and blacksmith shop were built, and tribal members did not adjust well to farm life. When the sturgeon were running, the would-be farmers tended to follow ancient pursuits rather than planting seeds or harvesting corn. Honorable and ancient ways made farming difficult.

The tribe's leadership vigorously protested the corruption of the Indian agents and the agent's relatives and hangers-on hired to implement the 1831 treaty. Some tribal members even made an effort to learn the new ways the government insisted they learn. But Washington failed to pay attention to Menominee protests, the desire of non-Indians for Menominee land became too great for the government to ignore, and the result was the 1848 Treaty of Lake Poygan, which took away all Menominee Wisconsin lands. The government still wanted the tribe to become farmers, but someplace other than where they had settled after the treaty. Menominee land was again needed for settlement.

By 1854 the effort to turn Menominee into farmers was again at the forefront of government efforts. At that point the Treaty of the Wolf River had made the tribe the owners of valuable reservation timberland. To Wisconsin's pine barons the Menominee were the government's wards, however, and thus did not control their homeland. The land, and thus the Menominee Forest, was held in trust. If the government could profit from selling Menominee timber, then the government could receive an entrepreneurial return for their Menominee investment. The pine barons would make sure the government would profit if only they would be allowed to earn their share of the profits to be made. The Menominee future was in farming, not timber.

In 1853 George W. Huebschmann was appointed to the Northern Superintendency, which had responsibility for the Menominee Reservation. Huebschmann was a true believer in the proposition that the Indian future was in farming, not timber. As soon as he was appointed he began working the reservation with great energy in an effort to make U.S. Indian policy work. For awhile he could even claim success. However, the long-term fate of the effort to make the Menominee farmers was determined as early as 1864 when a major crop failure caused Menominee starvation again (Beck 1994, 156). As has been noted, the Menominee Reservation is rocky, with a variety of poor farming soils. Although the effort to make the Menominee farmers continued into the 1970s and 1980s, the effort was doomed to failure soon after it had started.

The simple truth was that most Menominee not only had to struggle with the nature of reservation land in order to become farmers, they also had to struggle for what they believed was a better way of life. Farms followed the destruction of the forest, their home for centuries upon centuries of time. Farms were part of the destruction that led to the tribe's impoverishment and dependency. Farming also interrupted the seasonal occupations that had engaged Menominee from the distant past. From the Menominee point of view, farms ate up the land and ended the richness of the forest. They had heard the rasping saw whispers that destroyed the forests. They believed they could find a better way of living. Farmers were not the salt of the earth. Farmers made the land sore with their plows and the insatiable appetite of the logger's ax for trees and more trees.

What the Menominee wanted, they decided, was to be loggers. They were a woodland people. They had always lived off the forest's abundance. Loggers lived off the forest's abundance. If white loggers could make a living harvesting trees, the Menominee could do the same.

One indication such a Menominee endeavor could succeed was the amount of maple syrup being tapped, boiled down, and marketed from the reservation. In 1859 more than twenty thousand pounds of maple sugar was made, according to a report to the Commissioner of Indian Affairs. In 1863 the Menominee made forty tons (Hoffman 1896, 288). Maple syrup, combined with wild rice and berry gathering, provided both cash and subsistence to the tribe. Many Menominee, and especially Civil War veterans returning to the reservation after the war's end in 1865, believed they could do the work of logging as well as white loggers. After all, had they not just fought a war standing shoulder to shoulder with white American soldiers? Why should they not be allowed to harvest timber on their own lands, secured by treaty, and end the dependency, poverty, and starvation that was so much a part of reservation life?

The answer to this question was, of course, that the government held the reservation lands in trust and thus controlled Menominee lands. The Menominee could not cut down a single reservation tree without the Indian agent's approval. The Indian agents did not want the Menominee to be lumberjacks. They wanted them to be farmers. Powerful political forces also wanted Menominee timber for the profits it represented. Philetus Sawyer, a pine baron from Oshkosh, Wisconsin, who wanted the lumber company he owned to secure the Menominee franchise, was one of the U.S. Senate's most powerful members as the 1800s drew to a close.

The entire story of how the Menominee finally achieved the right to harvest their timber will be told in later chapters, but the point is that by the late 1800s the Menominee's historical experiences had led them to consider the Menominee Forest their most important asset. The forest was a link to the days when the Menominee had been a proud, independent people with a rich language, culture, spiritual life, and way of living. It also represented a way out of the morass of U.S. government farming programs and offered a life that seemed more inviting than that offered by cleared land and the plow. The Menominee could see themselves harvesting trees and sawing the harvest into lumber. They could see themselves working long, hard hours in the woods, working during the seasons when trees could be removed from the forest, and then hunting, fishing, gathering, and making family life strong and good when timber could not be harvested. In the midst of poverty, in the midst of changes that had fallen hard upon the tribe with the American ascension to power in Menominee country, they could see a prosperous, good future.

This future would be different from the future the Americans envisioned for them.. The Americans believed they should divide their lands so that each of the tribe's families could farm eighty or more acres. The white man believed in hurting the earth and ruining things. The Menominee would be different. They would own their land in the way their ancestors had owned the land. It would benefit no one individual nor be dedicated to the benefit of groups of individuals. It would serve all tribal members equally. No individual would own the land. The tribe would hold the land for all the people.

The Menominee would also not treat their forest the way the Americans treated the forests they had seized from the Menominee during the treaty period. They would cut the forest's dead and dying trees from the rising sun toward the setting sun until they had reached the end of reservation land. Then they would start the cycle of harvest all over again. They would cut the forest, but they would not destroy it. They would preserve it forever.

The first time the Menominee were allowed to harvest their forest, in 1854, the Indian agent was authorized by Washington, D.C., to allow a harvest on a year-by-year basis. The congressional act that made the harvesting possible also started laying the foundation for what the Menominee would later call sustained yield forestry. In 1908 the LaFollette Act (35 Stat. 51) allowed the Menominee to harvest their forest on an annual basis using the sustained yield practices that have been the heart of the tribe's forestry effort ever since.

Why This History Is Important to a Study of Menominee Sustainable Development

From this brief accounting of a section of Menominee history, a number of sustainable development "lessons" can be derived. The first of these is that the Menominee developed early ideas about sustainability out of loss, poverty, and degradation. They began the 1800s in control of an immense territory that provided them with food, clothing, shelter, and an independent way of life. During their journey through the century they lost nearly all their land base; their forest home largely disappeared from lands where they had once lived; their lives became dependent upon the American government that had taken so much from them; and they faced questions about their survival as a people. By the century's end what they had left was their forest and 240,000 acres. They mistrusted the idea they should be farmers and felt wrong about what the Indian agents were trying to accomplish.

They also knew, without doubt, that forests, and ways of life based upon forests, could die. Only if they acted could they hope to keep a forest and a forest people alive. They realized, partially as a result of the Treaty of Lake Poygan disaster, that they could lose all they had. If they were to preserve the forest and land they valued, they would have to do it for the long term, not the short term. If the Menominee Indian Reservation was all the land the tribe could have, then it would have to be preserved for their children and their children's children on into an unforeseeable future of generations.

By the end of the 1800s the tribe also realized the power of the American government and the white man's tools. They had developed early the habit of protest and complaint designed to frustrate the white man's designs. They had also learned the importance of going beyond the Indian agent to deal directly with the powers in Washington, D.C. The world's Philetus Sawyers could not be ignored. If the Menominee were to achieve what they were setting out to achieve, they had to work within the American political establishment.

Therefore, historically driven characteristics are important to the Menominee practice of sustainable development. These include the idea that sustainability was developed as part of a survival strategy. The Menominee did not accidentally happen upon the idea of sustaining their forest forever. They knew they had to find a way to make a living in the immediate future and also knew they had to plan for a period of time longer than could be imagined by any living person. The second key idea is that generational time is important. If you are going to sustain a forest or anything else worth sustaining, you have to develop ways to sustain it over an unseeable span of years.

The third idea is that political actions are important. The Menominee faced official corruption time and again, but they accepted nothing without protest and complaint. They sent delegations to Washington. They met incessantly with the Indian agent. They demanded what was theirs, learned how to read the treaties they had signed, and eventually learned about American jurisprudence—all in an effort to secure what they had and intended to maintain.

The last major idea the Menominee discovered was that they would have to make their own path in creating a sustainable life. They could not depend upon the American way. They were not farmers. Rather, they had to protect their land, its forest, and its water assets from the white man. They determined to own their land and its assets communally rather than individually.

Within this history is also the idea that sustainability is a powerful Menominee force partially because of tribal history. The Menominee sustain their forest and environment with emotion as well as with science and technology. Menominee sustainability has gone beyond a way of living and looking at the world. It is a protection, a song, a powerful sense of being that comes from a powerful history and powerful historical messages. In this case the message is, if you do not sustain the forest and environment, then you cannot sustain yourself, honor your ancestors, or provide for the children of your children into the future's unknowable reaches. Sustainability and your people's long-term well-being are one and the same thing.

4

Culture and the Spirit

An Exposition on the Menominee Sustainable Environment Ethic

MENOMINEE HISTORICAL EVENTS clearly provided the tribe with motivation in the 1800s to support and develop sustainable forest policies. Menominee culture has played a similar role in such policy development, but its contributions are difficult to identify.

The major reason this is true is that Menominee culture was in a state of crisis and change during the period when the first sustainable policies toward the Menominee Forest were developed and implemented. The cultural imperative on the reservation today is also cloudy. Certain cultural characteristics identify the Menominee as a people. A good example is the constant attention to prayer in the Menominee language that is part of nearly every reservation meeting and gathering. Songs and dancing are also powerful constants in Menominee life. No one can spend much time on the reservation without finding themselves spectators at a powwow, an honor song ceremony, or a Menominee prayer spoken by a tribal elder or religious leader. Yet, there was no general agreement about Menominee culture at the time of the first Menominee foresters, and no general agreement exists today.

Cultural differences characterize all North American tribal groups, even when the tribe is as culturally unified as the Navajo. In a 1991 article about social forestry on the Navajo Nation, the authors described a land use survey that illustrates the differences in contemporary Navajo culture. According to the authors,

> All land-users surveyed are "traditional" to some degree. All graze livestock and live in rustic log dwellings without electricity or run-

ning water. Most speak the Navajo language. However, the degree of traditionalism varies substantially. For example, some families cook with propane and some no longer practice the Navajo religion. . . . The traditional Navajo lifestyle is a pastoral one. The herd provides traditional Navajos with a sense of material and psychological security and is the basis for social organization of residence groups. All of the families interviewed graze livestock—primarily sheep and goats, but many graze cattle and horses too. Navajo families can no longer rely exclusively on their herds for income. The cultural importance of maintaining the herd may be even greater than its economic return. . . . Many areas in the forest are considered sacred by traditional Navajos. Most of these areas cannot be logged without destroying their spiritual values. Sacred areas include mountaintops, sweat lodges, burial grounds, old homesites, lightning-struck trees, and areas important for the collection of medicinal plants. . . ." (Einbender and Wood 1991, 15)

Neither the 1800s Menominee nor those of today could be described using such culturally specific language. Although a survey could probably be conducted on the reservation with "traditional" respondents, the sample would of necessity be small. Menominee spirituality, unlike that of the Navajo, is too diffuse, at least in this era, to hold that specific places are spiritual.

The reason the reservation lacks a dominant culture dates back to the 1600s. Not long after the first contact with Nicolet, Menominee relations with the French were cemented by marriages between early fur traders and Menominee wives. As Hoffman put it:

The relations between these French settlers and the natives were undoubtedly of an amiable character, as the general attitude and conduct of the French were rather of a conciliatory nature, whereby their representatives gained unusual confidence and good will among the natives—an attachment that was furthermore strengthened through the frequent selection by the French of Menomini wives. (Hoffman 1896, 16)

These marriages not only gave the Menominee a better understanding of European culture than was available to other tribes, but they also affected Menominee culture itself early in post-European-contact days. The children of these marriages usually, though not always, stayed with the tribe and practiced its ways. They also acted as intermediaries between first the French, then the British, and then the Americans. They played this role because of their fathers, who taught them French ways.

Aboriginal Menominee culture was also changed significantly by a historical event that occurred in 1637 when the Jesuit missionary Father Louis André converted several clans of the tribe to Christianity. Like many stories about Menominee culture, the Menominee conversion has a powerful magic. Father André described the event in a letter to his Jesuit superiors:

> When I arrived among them (the Folle Avoine, or Menominee) at the end of April, 1673, I gathered all the most notable persons, to inform them of my intention in visiting them. I also asked them what was meant by a picture of the sun that one of them had painted upon a piece of board. This picture was tied to the end of a pole, which was also painted in the brightest colors; and on this pole, at the height of a man, was suspended a sheaf of small cedar sticks, cut so as to serve as scats for the nets that are used in catching sturgeon, like the pieces of cork that are fasted to all kinds of nets in France. I therefore asked for what purpose they had set this sort of votive offering. They replied that it was a sacrifice—or rather, to use the proper expression in their language, "an exhortation"—which they had made to the sun, to entreat it to have pity upon them. As they believed that the sun was the master of life and of fishing, the dispenser of all things, they begged it to send the sturgeon into their river, and to make their fishing prosperous. They added they had long been expecting the sturgeon in their river and feared they would not come to it. In fact, they had reason to apprehend this, for the sturgeon had already entered the Pechetick river and that of Oukatoum, which are farther from the lake that is the river of the Maloumines. After disabusing them of the idea which they had of the sun and explaining to them in a few words the principal points of our Faith, I asked them whether they would consent to my removing the picture of the sun, and replacing it by the image of Jesus crucified. They replied, all together and repeatedly, that they consented; and that they believed that God was the master of all things. It was already late when they gave me this assurance of their good will; this did not prevent me from taking advantage of their favorable state of mind, and I put my crucifix in the place of the picture of the sun. On the following morning, sturgeon entered the river, in such great abundance that these poor people were delighted, and all said to me: "Now we see very well that the Spirit who has made all is the one who feeds us. Take courage; teach us to pray, so that we may never feel hunger." (André 1674, 275)

Not all the Menominee clans accepted Father André's miracle. Descendants of the Menominee who rejected the Catholic missionary still live in the reservation village of Zoar.

Still, the marriages between Menominee women and French men

and the conversion of significant numbers of Menominee to
Catholicism made Menominee culture complex soon after the first
contact with European culture. These events affected even traditional
Menominee culture.

By the time Hoffman was writing in 1892, the Menominee had
four major societies: (1) Mita'wit, or Grand Medicine Society, (2)
Tshe'o, or Juggler, (3) Wa'beno, or Men of the Dawn, and (4) Dreamers'
Society. The Jugglers possessed the most power, having the gift of
healing, but members of other societies also possessed important
powers that all traditional tribal members recognized. The Mita'wit
and the societies of Hoffman's time were a recent innovation imported
from the Anishinaabe and then changed to better reflect Menominee
tradition. By 1896, too, even these ceremonies were dying out. As
Hoffman explains:

> The Mita'wit . . . appears to receive less attention each year, the rea-
> sons being attributable to a variety of causes, chief among which are
> (1) the fact that many of the Indians are adopting Christian religion,
> as they begin to perceive the improved condition of those who have
> done so; (2) because many of the younger men are attending school,
> and begin to observe the futility and uselessness of the various
> dances; and (3) the old men and women mita'wok are slowly dying
> off, which makes it difficult to find candidates to fill their places
> (Hoffman 1896, 137–38).

In effect, the development of Menominee sustainable policies was
occurring during a time when traditional culture seemed to be dying.
Hoffman predicted, erroneously, that the culture would become more
of a tradition than a practice in a short period of time (Hoffman 1896,
138). Many Menominee leaders who championed forestry and the
practice of sustainability were also Christians and not traditional
Menominee, even though many traditional Menominee did work on
the logging crews.

By the early 1970s when George and Louise Spindler were studying the Menominee, the tribe's culture had become even more complex than in the late 1800s:

> The Menomini are not today a single people, even though Menomini
> distinguish themselves not only from Whites, Blacks, Chicanos, and
> other non-Indians but also from other Indians. Many forces have torn
> the Menomini apart. They have made several distinctive adaptations
> to the demands and threats of the dominant American culture and
> society. . . .

Though the Menomini made these adaptations over a period of time, and neither their adaptations nor the conditions to which they adapted remained constant, there are certain consequences that remain in force today. There are, within the Menomini community, five major cultural divisions. The *native-oriented* group maintains the Dream Dance, called the ni'mihe'twan (or, in Hoffman's language, the mita'wit) in the native language, the Medicine Lodge (the mete'wen), the War, or Chief's, Dance *(okeceta'we seman)*, and other rituals, all "religious" in character. . . . The *Peyotists* are members of the Native American Church. Most were raised in culturally conservative households, at least through early childhood, and all have had substantial contact with Whiteman culture. . . . The *transitionals* include individuals loosely joined together in informal groupings, such as drinking groups, people who are almost wholly isolated, and people who are striving toward fuller participation in the non-Menomini world. It is a heterogeneous category. . . . The *lower acculturated* are people who are overtly assimilated into Western culture and who are not members of traditional or nativistic groups. They differ from the *elite acculturated* in that their social and economic position is somewhat lower. The lower-status acculturated support themselves with jobs that, while by no means menial, are not management or professional positions. The elite acculturated can be regarded as people who have, in whiteman terms, "made it." Included among them are entrepreneurs, men and women with supervisory positions in the lumbermill or associated enterprises, and skilled white-collar workers. There are no doctors or lawyers in the group, though no doubt there soon will be. There are also a number of people, mostly young, who have left the home community and are residing in Milwaukee, Chicago, and other large cities. (Spindler and Spindler 1971, 4–5)

Contemporary Menominee society is, if anything, more complex than the society described by the Spindlers. Not only do members of all five cultural categories still exist on the reservation, but there exists a strong pan-Indian movement whose members embrace the national powwow circuit, and the beliefs common among those who participate from all of the nation's tribes. A growing educated class also embraces traditional, native-oriented ways. Each of these groups holds different sets of beliefs and has different levels of passion about its beliefs. No firm agreement across these groups defines what it means to be a Menominee in the contemporary world.

This complexity would seem to argue that Menominee culture has little to do with the sustainable policies so evident across the cultural division lines on the reservation and even among Menominee living in the nation's cities. The Spindlers argued in 1971, in effect, that no single Menominee culture existed even though the Menominee believed they were different from other Americans. But the evidence is that a strong cultural component is integral to Menominee sustainability. Menominee culture may be complex, but it is a powerful force among the Menominee both on and off the reservation.

The best place to look for evidence of the importance of Menominee culture to Menominee sustainable development is in the Menominee-generated literature about sustainable development. In the *Menominee Tribal Enterprises Forest Management Plan 1996–2005*, an unlikely source for comments about culture, the following statement is made:

> Unlike many non-Indian concepts of man and nature, the Menominee people do not view themselves as separate from the forest, or the forest and its creatures independent from them. The Menominee culture exists in harmony with Mother Earth, understanding the circle of life. The forest, properly treated, will sustain the Tribe with economic, cultural, and spiritual values today and for future generations. This has been taught and practiced on Menominee lands for more than 8,000 years, and accounts for the quality of the forest on the Reservation today. (Menominee Forestry Department 1995, 110)

A Menominee Tribal Enterprises sales brochure stresses the importance of culture to the Menominee's sustainable forestry ethic:

> Early records indicate the Menominee as living in villages at the mouth of the Menominee River, and it was here the tribe had its beginning. According to a sacred origin myth of the Menominee, the Great Bear emerged from the ground, took human form and was made an Indian by the Great Mystery. Being alone he called to himself an Eagle flying high above to be his brother and descending took the form of a human and an Indian. As the two journeyed up the river they met a beaver and made him their brother, as were the Wolf, Sturgeon, Crane, Moose and White-tailed deer being made their brothers. Other animals, fish and birds were made their brothers forming the clans of the tribe and the first Menominees.
>
> Today, the Menominee live in an area which is part of their original domain. This pristine environment with clear running streams, sparkling lakes and acres of forested land are filled with animals, fish and birds like it was when the creator made the first Menominee. (Menominee Tribal Enterprises Brochure 1994)

When asked about why the Menominee have preserved their forest and environment for so long, Menominee people, irrespective of the cultural group the Spindlers would have placed them in, invariably respond with a reference to elders at the time when the reservation was first formed and a description of the Menominee's spiritual and cultural nature.

If Menominee culture is so fragmented as to hardly qualify as a single culture, then why is the belief in the beneficence of Menominee culture so prevalent today? And if Menominee culture and spirituality are important to the way the tribe has sustained their forest and environment, then what cultural elements have led to the ethic behind Menominee decision-making about forest and environmental policy? As has been noted, answering these questions is not an easy task.

Where the Menominee Came From

The answer to these questions lies in, first, the definition of culture, and then in the way Menominee life fits that definition. Culture, according to the *Funk and Wagnall Encyclopedia*, "is the sum total of all contributions of a group of people, in a designated area, within a given time. It represents, more specifically, the aesthetic or intellectual achievement or appreciation of an individual or a society, and also the lifestyle of a society as passed on from generation to generation" (Future Vision Multimedia, Inc. 1995). Using this definition, Menominee culture, based upon the prisms provided by Menominee stories, becomes much more concrete. The Spindlers were accurate in their depiction of an acculturating society in the midst of transformation forced by "Whiteman's culture," but their description tells only part of the Menominee story. All Menominee share a history, legendary and religious beliefs, an oral tradition, the Menominee Forest, the way they have managed the forest, an environment, stories, a political system, a constitution, and a heritage in common. They often disagree about the significance of any one of these elements. Divisions on the reservation often rise out of these fundamental disagreements. But, as complex as it is, Menominee culture is as rich and full of import as any other culture found on earth. Aesthetic, intellectual achievement and an appreciation of individuals and society give the culture a powerful life. The lifestyle of Menominee society as it is being passed from generation to generation also makes the Menominee unique as a people.

Menominee culture is central to understanding Menominee sustainable development because the culture, and thus the Menominee story, has at its heart the forest. As Nicholas C. Peroff, a prominent historian, has pointed out, "from the earliest archaeological traces of the tribe, dating back over five thousand years, the forest had remained central to the lives of the Menominee Indians" (Peroff 1982, 3). The continuation of forest culture, in the midst of historical upheaval, explains the Menominee insistence that their culture is at the heart of their forestry and manufacturing practices and provides the center that holds the tribe together. In a real sense Menominee culture is built upon the theorem that the Menominee Forest not only can, but will, last forever. Without their forest, the Menominee would lose a key component of their identity.

This is not to say the cultural complexity described above is any less complex. Menominee beliefs are, when analyzed, always complex. For instance, most Menominee, when asked, will express a strong affinity for the earth. Mother Earth, the Great Spirit, and the Menominee Forest are thanked again and again on the reservation in prayers for providing for the tribe in a multitude of ways. This sense of affinity rises from the story of the Menominee people's origin:

> When the Great Mystery made the earth, he created also numerous beings termed manidos or spirits, giving them the forms of animals and birds. Most of the former were malevolent ana'maqki'u (underground beings); the latter consisted of eagles and hawks, known as the Thunderers, chief of which was the Invisible Thunder, though represented by Kine'u, the Golden Eagle.
>
> When Masha' Ma'nido—the Good Mystery—saw that the bear was still an animal, he determined to allow him to change his form. The Bear, still known as Nanopke, was pleased at what the Good Mystery was going to grant him, and he was made an Indian, though with a light skin. This took place at Mi'nika'ni se'pe (Menomini River), near the spot where its waters empty into Green Bay, and at this place also the Bear first came out of the ground. He found himself alone, and decided to call to himself Kine'u, the Eagle, and said, "Eagle, come to me and be my brother." Thereupon the Eagle descended, and also took the form of the human being. While they were considering whom to call upon to join them, they perceived a beaver approaching. The Beaver requested to be taken into the totem of the Thunderer, but, being a woman, was called Nama'kukiu' (Beaver woman), and was adopted as the younger brother of the Thunderer. . . . Soon afterward, as the Bear and the Eagle stood on the

banks of a river, they saw a stranger, the Sturgeon (Noma'eu), who was adopted by the Bear as a younger brother and servant. In like manner Omas'kos, the Elk, was accepted by the Thunderer as a younger borther and water-carrier.

At another time the Bear was going up Wisconsin river, and becoming fatigued sat down to rest. Near by was a waterfall, from beneath which emerged Moqwai'o, the Wolf, who approached and asked the Bear why he had wandered to that place. The Bear said that he was on his way to the source of the river, but being fatigued and unable to travel farther, he had come there to rest. At that moment Ota'tshia (the crane) was flying by, when the Bear called to him and said: "Crane, carry me to my people at the head of the river, and I will take you for my younger brother." As the Crane was taking the Bear on his back, the Wolf called out to the Bear, saying, "Bear, take me also as a younger brother, for I am alone." The Bear answered, "Come with me Wolf, and I will accept you also as my younger brother." This is how the Crane and the Wolf became younger brothers of the Bear; but as Moqwai'o, the Wolf, afterward permitted Anain', the Dog, and Aba'shush, the Deer, to join him, these three are now recognized as a phratry, the Wolf still being entitled to a seat in council on the north side and with the Bear phratry. (Hoffman 1896, 39–40)

Still, the story's importance does not mean it is literally believed by all Menominee. Such is not the case. A strong Catholic, part of what the Spindlers described as the "acculturated elite," would not for a moment consider this story the true story of the Menominee origin. On the other hand, reservation traditionalists do believe this is how the Menominee came to be.

What makes the origin story important is that it is a touchstone for all the Menominee people, whatever their religious beliefs. The story's motifs are repeated over and again by Menominee artists in murals and sculpture. Menominee storytellers still speak the ancient words, burning tobacco before the story, to make the story acceptable to the Great Mystery.

The Bear was the first Menominee. The Bear came from underneath the earth and is symbolic of the earth. The other traditional clan leaders, from the Eagle to the Beaver Woman to the Wolf, all became brothers to the Bear—to earth. Thus, the Menominee have a special relationship to the earth. Almost without exception tribal members see themselves, even if their circumstances are unpleasant and destructive, as a people who protect the earth, the forest, and its creatures. If they fail in this duty, then they have failed as Indians, as human beings.

Ancient Menominee culture is not important because of its contemporary adherents—although those adherents play an important role in Menominee life—but because its traditions are still living among the Menominee people in ways adapted to contemporary life. The historian Robert E. Bieder points out that in the time before the Europeans, "religion was integral to Menominee community life and essential to its well-being." The rituals and daily life of the Menominee were animated by the tribe's religion:

> Both animate and inanimate objects in their world possessed spirits that had to be propitiated and manipulated through proper ritual procedures. Each animal had a spirit that had to be treated with proper respect. Failure to do so would prevent successful hunts. Even tools, like weapons and pots, possessed spirits that demanded proper attention in order to perform in useful ways. Luck in hunting or fishing did not exist. Humans were successful only if they had power or the proper medicine to control the numerous elements in their world. Humans survived only if the spirits helped them. (Bieder 1995, 29)

Many elements of this faith are no longer believed by the majority of Menominee, although, again, the reservation's traditional community and its influence should not be forgotten. But Menominee spirituality is a powerful part of Menominee community. Americans may presume to protect religious freedom by denying the right of prayers at public events, but at every Menominee Legislature meeting, every major committee meeting, every school event, and every community event, prayers are offered to the Great Spirit, usually, but not always, in the Menominee language. Those living and working on the reservation are in this way continually reminded of the spirit's importance to life. Practicing Catholics and Presbyterians participate in these prayers as fervently as members of the Native American Church and the traditional people. Menominee may live their daily lives in the same manner as non-Indians, even placing a satellite dish in their backyards, but the Menominee are still a spiritual people who display their spirituality, with great solemnity, on a regular, public basis.

So many aspects of the old religion are reflected in contemporary Menominee life that a substantial book would have to be written to describe the phenomenon. According to Alanson B. Skinner, one of the most knowledgeable scholars who has studied Menominee spiritual beliefs:

> The earth is believed to be an island, floating in an illimitable ocean, separating the two halves of the universe into an upper and a lower portion, regarded as the above and below of the benevolent and malevolent powers, respectively. Each portion is divided into four superimposed tiers, inhabited by supernatural beings, the power of whom increases in ratio to their remoteness from the earth. (Skinner 1921, 29)

No individual person could own the island earth. No individual could own any part of the island. The island had been made by the Great Mystery, and it was populated by greater and lesser spirits. This means, using Vine Deloria's words, that:

> From the days of the earliest treaties, Indians were shocked at the white man's attitude toward land. The tribal elders laughed contemptuously at the idea that a man could sell land. "Why not sell the air we breathe, the water we drink, the animals we hunt?" some replied. It was ludicrous to Indians that people would consider land as a commodity that could be owned by one man. The land, they would answer, supports all life. It is given to all people. No one has a superior claim to exclusive use of land, much less does anyone have the right to fence off a portion and deny others its use. (Deloria 1970, 82)[1]

Individual Menominee own land today, but the Menominee Forest has been held as tribal, rather than individual, property ever since the Menominee canoed up the Wolf River to their permanent reservation home after signing the Treaty of the Wolf River. One tenet of Menominee sustainable development stresses the importance of maintaining communal ownership of the forest and the Menominee environment. The people who came from the Bear and the Bear's brothers are responsible for protecting the earth. This is not an individual responsibility, but is the responsibility of every individual. No individual or group of individuals has the right to destroy any part of the island where spirits and Indians (and these days non-Indians) live.

The current draft of the "Forest Management Plan" does not discuss issues such as wildlife, cultural resources, and endangered resources (Menominee Forestry Department 1995, 17–19) simply out of an awareness of the contemporary environmental movement and the science of ecology. This is a highly technical document that details the goals, objectives, and methods Menominee Tribal Enterprises intends to pursue and use in forest management through 2005. Its policies are described in detail in chapter 8. But it is also a document that allows the Menominee to confirm they are still Menominee even after their association with the Long Knives. To the writers of the plan, each animal has a spirit that has to be treated with proper respect. Both animate and inanimate objects possess spirits that need protection if the Menominee and earth are to remain healthy and whole. The cultural remains of the old ones found in the forest must be preserved and protected. No holy sites mark the forest in the Navajo tradition, but the legacy left by the reservation's original inhabitants are spiritually and culturally important. This legacy must be protected and preserved.

What Menominee culture creates is a powerful environmental, sustainable ethic. This ethic is made even more powerful, at least in Menominee thought, by the events and pain of the losses experienced with the American ascendancy over tribal land. It is made more powerful yet by a forest that has sustained them as a people for more than five thousand years.

A Sustainable Environmental Ethic

The Menominee sustainable environmental ethic, as has been indicated, comes from multiple sources. These include historical, cultural, ancestral, and religious/spiritual sources. They also include the tension that exists on the reservation between Indian and non-Indian worlds. Even Menominee economics plays a role in defining what the ethic is and how it works. So far this chapter has shown how elements of early Menominee culture and spirituality have become part of the overall Menominee approach to sustainability. It has also established that even though Menominee culture is complex, its aesthetic is rooted in pre-European Menominee intellectual achievements, especially as related to the Menominee environment, forest, and lifestyle passed from generation to generation.

This examination of culture and spirituality leads to questions about the ethic that leads a people to develop sustainable development policies strong enough to weather the vicissitudes of time. An ethic is a set of principles, or standards, governing human conduct. Menominee history, culture, and spirituality have been essential elements in creating the Menominee ethic. They have provided the frame that makes the Menominee a people and has allowed the tribe to act in ways substantially different from the actions of the larger American society.

The principles that exist, like Menominee culture itself, are not always obvious. This book cannot identify all these principles. Nor do all the principles explored in this book appear in this chapter since the chapter's focus is the examination of principles rising from Menominee culture and spirituality.

The poet, essayist, and deep ecologist Wendell Berry provides as good an introduction to the Menominee ethic as is available. Berry visited the reservation in 1994, and his thoughts about his visit were recorded in an essay entitled "Conserving Forest Communities."

> The paramount lesson (learned from the Menominee) undoubtedly is that the Menominee forest economy is as successful as it is because it is not understood primarily as an economy. Everybody I talked to on my visit urged me to understand that the forest is the basis of a culture, and that the unrelenting cultural imperative has been to keep the forest intact—to preserve its productivity and the diversity of its trees, both in species and in age. The goal has always been a diverse, old, healthy, beautiful, productive community-supporting forest that is home, not only to its wild inhabitants, but to its human community. To secure this goal, the Menominee, following the dictates of their culture, have always done their work bearing in mind the needs of the seventh generation of their decendants. (Berry 1994, 19–20)

Marshall Pecore, head Menominee forester, expressed the ethic in an even more powerful way in an article he wrote for the *Journal of Forestry*. After explaining that the question of "whether and how to develop natural resources is a subject of great debate in Indian communities, with the Menominee being no exception," he goes on to say that "there is broad agreement, however, that natural resource development must be part of a larger plan of social and economic development rather than an end in itself."

> It is said of the Menominee people that the sacredness of the land is their very body, the values of the culture are their very soul, and the water is their very blood. It is obvious, then, that the forest and its living creatures can be viewed as food for their existence. The Menominee people have balanced the concept of long-term, sustained-yield forest management with the shorter-term, diverse considerations of community stability and economic development. Their story is one of successful equilibrium between harvesting and using only what the land can provide, and maximizing the jobs and other

economic benefits that flow from a sustained-yield harvest. (Pecore 1992, 12)

These two passages identify several major principles of the Menominee sustainable ethic. The first is that the forest is a resource to be used for the Menominee people's benefit. "The goal," according to Berry, "has always been a diverse, old, healthy, beautiful, productive community-supporting forest that is home, not only to its wild inhabitants, but to its human community." On the Menominee Indian Reservation community rises out of the forest's productivity. The forest is not to be protected and preserved as a wilderness that provides solace to the troubled modern soul. "The forest and its living creatures," according to Pecore, "can be viewed as food for their [the Menominee people's] existence." In fact, the major reason for maintaining "a diverse, old, healthy, beautiful" forest is so that the human, plant, and wildlife communites can survive and prosper together.

A second major principle is that Menominee culture and spirituality is not based primarily on humans as individuals. "It is said of the Menominee people," according to Pecore "that the sacredness of the land is their very body, the values of the culture are their very soul, and the water is their very blood." From Berry's non-Indian viewpoint, "the forest is the basis of a culture," and this fact makes the forest as a natural resource much more than an economic opportunity to be exploited. It weaves a bond between humans and the earth in a manner that protects the forest. If the forest is destroyed, then the culture—the very body, soul, and blood of the people—is also destroyed.

The third principle has to do with the sustainability of human community. Pecore notes that forest use is supposed to yield the maximum number of jobs and other economic benefits, but this maximum has to be balanced against "the concept of long-term, sustained-yield forest management." Berry, using a phrase more common to the neighboring Oneida and Stockbridge-Munsee tribes than to the Menominee, says that to secure forest productivity and economic benefit, "the Menominee, following the dictates of their culture, have always done their work bearing in mind the needs of the seventh generation of their descendants."

Joseph Frechette, quoted in the first chapter, outlined the idea of sustained-yield forestry by describing how the Menominee were to harvest the forest in cycles. "When you reach the end of the reservation," he said. "turn and cut from the setting sun to the rising sun and the trees will last forever."

This expresses the ethic's fourth major principle. Sustainability has an endless time horizon. An assumption is made that the Menominee and the forest will be related into the most distant of times. This relationship forces the Menominee to act toward the forest in ways that protect the forest forever.

Dr. Verna Fowler, the Menominee president of the College of the Menominee Nation, in a conversation occurring in an airplane flying from Detroit to Washington, D.C., explained the ethic in a different way:

> You have to realize that the Menominee only have one home, the reservation. Their ancestors going back untold generations are buried there, sometimes in places that are a complete surprise. Their children and their children's children and all the children of the generations to follow will have to live in the Menominee homeland. That means that the reservation is going to have to take care of all the generations that have been and all of those who will follow forever. (Fowler, conversation 7 May 1994)

The ethic is not only concerned with preservation of the forest, wildlife, and forest ecologies, but it is also concerned with human generations. Today's Menominee have the right to benefit from the forest as long as they protect forever the integrity of what makes it a forest. But in managing these two tasks, the Menominee also have to think of their continuity as a people stretching backward five thousand years and forward into the unforeseeable future. Culture arises from the forest, but it also arises from the Menominee generations.

Larry Waukau, the current Menominee Tribal Enterprises president, said, in an interview with the *Shawano Leader*, the daily newspaper of the small city closest to the reservation, that:

> The tribe's land ethic and management philosophy has always contained the three components of a sustainable system. First . . . [the system] must be sustainable for future generations. Second, the forest must be cared for properly to provide for the needs of the people. And third, we keep all the pieces of the forest to maintain diversity. (*Shawano Leader* 1995, 1)

This culturally based ethic has persisted for three centuries of Menominee contact with first European, then American society, evolving in sophistication and power over the years. The ethic has several parts, some of which may not have been described. Menominee culture is complex and often difficult to understand. But whatever its attributes, this ethic appears strong enough to continue forever into the future it speaks of so often.

5

Menominee Politics, Political Character, and Institutions

The Menominee Battle against Assimilation and How It Has Shaped Sustainable Development Efforts and Culture on the Reservation

THIS CHAPTER STARTS with a discussion of "the old time Menominee" (Keesing 1987, 18) and the political and institutional structures in place before the Europeans came. Although most of these structures' elements are no longer relevant, others have lived through time and affect the tribe's thinking, politics, and institutions today. The forces of assimilation and termination are then discussed, along with the social, political, and institutional effect of these forces. This part of the chapter explains much of the complexity of tribal politics today and also establishes the framework for understanding how the social, political, and institutional complexity of Menominee politics has helped to institutionalize sustainable development policies.

What has to be understood first, however, is that the Menominee see the universe in a much different way than non-Indians. Just how different these views can be is illustrated by a conversation held on 1 December 1913 between Edward E. Ayer, a government official sent by the Bureau of Indian Affairs to investigate tribal affairs, and a Menominee civil war veteran—a pagan in Mr. Ayer's terms—Mr. Wyeskesit. Mr. Wyeskesit started out the conversation by telling Ayer why he was there:

> The reason why I came to see this gentleman (Ayer) is to tell him
> how the Menominees are in poor circumstances. You see this city

(Neopit) here, it looks nice and good. But where I live is the poorest settlement that there is on the reservation—Zoa settlement, six miles away.

MR. AYER: What is the nature of the land up there; is there any good farm land?

WYESKESIT: A good land, timber land; but how am I going to use the farm?

MR. AYER: I am going to recommend that the Government will advance money enough to those who want to farm to build a house and barn and get a good team. Won't that help you?

WYESKESIT: I want the views of this proposition placed before the tribe, and whatever action my people will take, then I will give my opinion.

MR. AYER:

Q. I want your individual opinion.

A. My opinion is that when a man comes here I like to have him go and be my witness to the poor conditions we are placed in up there. We are in such poor circumstances we have no lumber; some of my people up there have bark for their roofs. We have no money to buy this lumber up here.

Q. How do they live?

A. Go around and hunt and trap.

Q. Is there much game on the reservation?

A. There is, but there are some us that are unable to hunt. We are old and sick.

Q. How old are you?

A. 72.

WYESKESIT: That was why I went to the war (the Civil War) to fight for this country so that the white people that I helped would help me when I am poor.

MR. AYER:

Q. How much pension do you get?

A. $208.00 a year.

Q. How much annuity do you get a year?

A. In the beginning I got $30.00 twice a year; but now the money is being used for other purposes; it is sunk here (in Neopit, at the sawmill).

MR. AYER: In the past two years $444,000 was deposited with the United States Government out of the profits of this plant. Isn't that good?

WYESKESIT: If this plant (the sawmill) was a paying proposition would I be poor; would I be hungry? The white people you see here, they are the people who have good things to eat.

MR. AYER: But over and above the entire expense of this plant for two years they have deposited $444,000 in the tribal funds at Washington. The money is your funds at Washington now amounts to about two million dollars. If this plant keeps on as it has for the last two years it will pay in a short time all the money that has been put into it.

WYESKESIT: Well, where is this money?

(Ayer 1914, 83–84)

This conversation illustrates that putting political or institutional concerns in non-Indian terms from a non-Indian viewpoint is dangerous. Ayer is trying to establish a rational American conversation with Wyeskesit. From Ayer's viewpoint the question is one of economics. Is the land around Zoa (Zoar) any good for farming? What do you think of my plan to provide resources for more Indians to get into farming? Is not the amount of money the Neopit sawmill has made during the last two years a good thing? Look at the money that has been built up in Washington for the tribe. Is this not a good thing?

Wyeskesit's answers come from a different political and cultural center than Ayer's questions. The land is good timber land, "But how am I going to use the farm?" Wyeskesit asks. Today's Menominee would respond in much the same fashion. Farming is okay for white people. The Menominee are a woodland people. The forest, irrespective of its financial benefits, has a benefit. The Menominee have no use for farming.

Wyeskesit is also concerned that his government visitor understand his reality. All that Menominee money in Washington is worthless as far as he can tell. The mill may be worthless. The people in Zoar have bark for roofs and not enough to eat. Only the white man has use of the money earned from the Menominee Forest. They eat good food while the Indian starves.

Besides, the white man is hard to understand. Wyeskesit fought in the Civil War, helping the white man during his time of need, but the white man has not reciprocated. He has not even come to Zoar to find out how poor the Indian people are. Wyeskesit is, of course, a reasonable man. He will consider Ayer's suggestion of providing money to those tribal members who want to farm, but only in the tribal context, not in the context of Wyeskesit as an individual. "I want the views of this proposition placed before the tribe, and whatever action my people take, then I will give my opinion." Ayer's ideas have no real meaning to Wyeskesit until his people have decided to take action upon it. When such action has been taken, Wyeskesit will then comment on the action, not the idea itself. Only when the tribe has acted is Wyeskesit willing to form an opinion.

The political scientist Deborah A. Stone believes that American politics are not centered in economic rationality,[1] as most policy analysts and political philosophers believe, but within the paradoxes of conflicting ideas. These paradoxes arise from conflicts inherent in the major political and social values held by most Americans (Stone, 1988). In the Menominee context, Stone's ideas help clarify some of the ideas that follow. Wyeskesit's political and social values come from a different center than those held by Ayer and non-Indian society. Menominee political conflict, however, deals with not only paradoxes inherent in Menominee society and values, but also with paradoxes inherent in American society and values. These paradoxes and their inherent conflicts create the reservation's political vortex. A description of these paradoxes and the political and institutional effects created by the vortex takes up most of this chapter. This chapter also attempts to explain how the institutions that maintain sustainability have resulted from this vortex. A suggestion is made that the complications arising from the political and institutional complexities that bedevil the reservation, paradoxically, serve the cause of Menominee sustainability.

The Old-Time Menominee and Their Institutions

Not much is really known about how the pre-European Menominee organized their society, but what is known is that organization mostly resulted from a mixture of religious and survival concerns. The pre-European period is important to an examination of Menominee sustainability because some of the tribal traits that make Menominee institutions and politics so difficult for outsiders to understand stem from early Menominee society.

The earliest Menominee probably lived, primarily, in two villages. Hoffman reports this in the story of how the Menominee came to be. "Ina'maqki'u (the Big Thunder) lived at Winnebago Lake, near Fond du Lac, Wisconsin. The Good Mystery made the Thunderers the laborers, and to be of benefit to the whole world." Later on the Thunderers decided to visit the Bear Village at Mi'nika'ni. When they arrived they asked the Bear to join them, promising corn and fire in return for rice, the property of the Bear and Sturgeon, which abounded along the Mi'nika'ni's waters. After being asked:

> The Bear family agreed to this, and since that time the two families have therefore lived together. The Bear family occupies the eastern

> side of the council, while the Thunderers sit on the western side. The
> latter are the war chiefs and have charge of the lighting of the fire.
> (Hoffman 1896, 40)

The significance of the two villages, and later on, one village, is that this indicates the Menominee were stable rather than nomadic. They grew crops, such as corn, harvested wild rice, hunted, tapped syrup from maple trees, and gathered blackberries and other northern fruits and berries. Less stable, more nomadic tribes depended on hunting for subsistence. Archaeological evidence indicates that the Menominee ranged throughout Wisconsin in their search for good hunting during certain seasons, but always returned home. The Thunderer clan provided leadership in wartime. The Bear clan provided civil leadership.

This moiety organization has extended into the present day in at least one way. When the Menominee settled the reservation, they tended to settle according to clan membership. Thus, although clan homogeneity has no presence on the reservation today, different clan families still live on different parts of the reservation.

Within each of the moieties were a number of phratries, or bands. Each phratry was related in Menominee religion with spirit beings of the earth, air, and water that formed the original tribe. As Keesing points out, the list of pre-European clans has been lost to history. Early documents list different phratries. In the origin story two are mentioned when the Wolf permitted the Dog and the Deer to join him as a human being. Other authors mention the White Beaver, the Wolf, the Turtle, the Crane, and the Bear as the principle clans with lesser phratries, including spirit beings such as the Turkey. Clan and phratry membership near the time when the Menominee met the French was set by patrilineal descent, although earlier in history matrilineal descent was important (Keesing 1987 37).

In village life clans and phratries provided leaders who made up the village council. The village council was responsible for dealing with non-tribal members, settling disputes, and managing tribal affairs. There were also war chiefs

> who won prestige through individual dreams or because of their
> prowess, though from such positions the civil leaders were not
> barred. These persons acted as keepers of the war medicines, as pub-
> lic spokesmen for the hereditary leaders, as masters of ceremonies
> during public celebrations, and as "police" and guardians of the wild
> rice at harvest time. (Keesing 1987, 40)

War was not common among the Menominee, however. The Menominee were a small tribe and usually negotiated with other tribes. They seldom attempted to force their will upon other peoples. When war occurred it tended to be "a matter of small parties and sudden actions." Young men usually tried to win reputations as warriors (Keesing 1987, 40), and the Menominee record as fighters has always been notable, but the tribe was not a warrior tribe like the Winnebago or Sioux.

Early Jesuit priests and explorers described the Menominee as a harmonious people with a surprising equality between chiefs and other tribal members. Tribal life was based on merit rather than on heredity or political influence. The early French considered the tribe hospitable and noted that disputes among tribal members were handled with moderation and forbearance. The entire village took interest in events wherein one individual committed some wrong or if a family suffered a tragedy. Keesing explained the tenor of early Menominee life in this way:

> As with innumerable small groups the world over the force of public opinion, and the benefits to be gained by adherence to the rules surrounding economic, social, ceremonial and like activities, these backed in many cases by supernatural sanctions, made for relatively harmonious community living. A "code of blood vengeance" is frequently referred to in the tribal lore and mentioned in some of the early records, and this doubtless came into play after inter-tribal or inter-group killings in order to even up unbalanced scores. . . ." (Keesing 1987, 41–42)

Some of these tribal characteristics are not only present today, but have been key to the establishment of sustainable development policies. They have also been important to the maintenance of those policies and the sustainable programs resulting from them. Social group cohesion has allowed the Menominee to differentiate themselves from non-Indian society and helped them to act differently from the society surrounding them. As will be seen later, the ability of tribal members to stand on an equal footing with tribal leaders and move into leadership positions as a result of an ability to excel has also played an important role. Even the council form of government, with its emphasis upon the power of oratory and consensus, has been important to what the Menominee have achieved with their forest and environment.

From this early point forward, the clan and phratry structure faced severe challenges. A hallmark of Menominee culture is its ability to adjust itself to the larger, ever-changing American culture while maintaining a core identity. The French, British, and Americans all forced changes upon Menominee politics and institutions. Still, some of the tribe's key political and institutional traits extend from its earliest history.

A Changing Political and Institutional Structure

The first post-European changes in the tribe's political relationships and institutional structure happened after the French introduced the fur trade to the Upper Great Lakes region. With the fur trade's introduction two major needs imposed themselves upon Menominee life. The first had to do with the trade's nature. To trap beaver that brought the benefit of French goods, Menominee had to leave their village and range far afield. As time went on and beaver became increasingly scarce, Menominee trappers ranged farther and farther afield in competition with other tribal groups. The whole tribe could not venture forth on these deep woods excursions. Rather, the tendency was to set forth in independent bands, or small groups. Menominee historians have debated about whether or not the band structure predated the French, but the point is that the band structure grew in importance with the fur trade's inception.

Normally these bands were extended family members that could work efficiently together in the effort to find and trap as many beaver as possible. The fur trade's major effect was that it made village life less settled and started the Menominee on their long journey of adaptation and change. The band system also tended to strengthen the family's importance in the tribe and reduced, at first to a minor degree, clan and phratry importance. This does not mean that the clan and phratry structure disappeared. That would not happen until long after the Americans had achieved political ascendancy. The clan and phratry structure stemmed from Menominee religion and is still important on the reservation today to both traditional and non-traditional Menominee. But the increasing importance and reliance on families that started during the fur trade period would, later on, during the American period, center Menominee politics in extended family groupings.

The second major need created by the fur trade was for tribal members (or relations after the French started marrying Menominee women) who could successfully deal with non-Menominee society.

The French, at the time of first contact, were both strange and power-ful. Hoffman tells the Menominee version of just how strange the French were, as recounted to him by a tribal elder, Waios'kasit:

> When the Menomini lived on the shore of the sea, they one day were looking out across the water and observed some large vessels, which were near to them and wonderful to behold. Suddenly there was a terrific explosion, as of thunder, which startled the people greatly.
>
> When the vessels approached the shore, men with light-colored skin landed. Most of them had hair on their faces, and they carried on their shoulders heavy sticks ornamented with shining metal. As the strangers came toward the Indians the latter believed the leader to be a great ma'nido (Spirit), with his companions.
>
> It is customary, when offering tobacco to a ma'nido, to throw it into the fire, that the fumes may ascend to him and that he may be inclined to grant their request; but as this light-skin ma'nido came in person the chief took some tobacco and rubbed it on his forehead. The strangers appeared to be desirous of making friends with the Indians, and all sat on the ground and smoked. Then some of the strangers brought from the vessel some parcels which contained a liquid, of which they drank, finally offering some to the Menomini. The Indians, however, were afraid to drink such a pungent liquid indiscriminately, fearing it would kill them; therefore four useless old men were selected to drink the liquid, and thus be experimented on, that it might be found whether the liquid would kill them or not.
>
> The men drank the liquid, and, although they had previously been very silent and gloomy, they now began to talk and to grow amused. Their speech flowed more and more freely, while the remainder of the Indians said, "See, now it is beginning to take effect!" Presently the men dropped down and became unconscious; then the Indians said to one another, "Now they are dead; see what we escaped by not drinking the liquid!" There were sullen looks directed toward the strangers, and murmurings of destroying them for the supposed treachery were heard.
>
> Before things came to a dangerous pass, however, the four old men got up, rubbed their eyes, and approached their kindred, say-ing, "The liquor is good, and we have felt happy; you must try it too." Notwithstanding the rest of the tribe were afraid to drink it then, they recalled the strangers, who were about to return to their boats.
>
> The chief of the strangers next gave the Indians some flour, but they did not know what to do with it. The white chief then showed the Indians some biscuits, and told them how they were baked. When that was over, one of the white men presented to an Indian a gun, after firing it to show how far away anything could be killed. The Indian was afraid to shoot it, fearing the gun would knock him

over, but the stranger showed the Indian how to hold it and to point it at a mark; then pulling the trigger, it made a terrific noise, but did not harm the Indian at all, as he had expected. Some of the Indians then accepted guns from the white strangers.

Next the white chief brought out some kettles and showed the Indians how to boil water in them. But the kettles were too large and too heavy to carry about, so the Indians asked that they be given small ones—cups as large as a clinched fist, for they believed they would grow to be large ones by and by.

The Indians received some small cups, as they desired, when the strangers took their departure. But the cups never grew to be kettles. (Hoffman 1896, 214–16)

To the Menominee these new people were ma'nido, spirits like the light-colored Bear that came out of the ground and started Menominee history. These generous spirits gave the Indians marvelous things. But these marvelous things turned out to be part of the greater plan to make the Menominee dependent upon the French. Guns could be given away, but only the French could provide ammunition. A taste for liquor and flour could be created, but only the French could provide liquor and flour. The gifts had a hidden hook in them. They linked the French and Menominee together in ways that would eventually lead to American ascendancy, the loss of Menominee land, and the creation of Menominee sustainable development.

In time the Menominee learned of the French plan. They integrated their lives into the worldwide economy served by the furs the French wanted the Menominee to trap. Still, those who could deal successfully with the French, the British, and then the American strangeness became important to the tribe. Such tribal members were often distrusted since the white man's values and attitudes were sometimes contagious. They changed the way people related to Mother Earth and the Great Mystery and the clans themselves. But in the end the Menominee learned how to deal with the Europeans, and those relation who could wisely advise the tribe had a value analogous to the value of traditional civil and war chiefs.

From the French to the British

As French influence declined in Menominee territory and the British gained ascendancy, the tribe's political character changed. These changes helped set patterns the Menominee have followed to the current day, especially when interacting with the more powerful societies with which the Menominee have had to coexist.

The fur trader Nicholas Perrot, one of the Frenchmen with the longest Menominee association, described the change in this way:

> At the present time, it is evident that these savages [who have intercourse with the French] are fully as selfish and avaricious as formerly they were hospitable. (Keesing 1987, 41)

Another way of describing this change, using modern vernacular, is that the Menominee wised up. They became hard bargainers and difficult-to-please customers. They began to distrust the white man, even the French, whose men often became husbands of Menominee wives, and tried to make sure that they, rather than the white man, got the best end in the increasingly frequent dealings between the two races.

The second major change happened when the French gave way to the English. This was a hard period for the tribe, though nowhere near as hard as the times to follow. In spite of their "selfishness" and "avariciousness," the tribe had been good French allies. Not only were an increasing number of children of mixed Menominee and French heritage, but the Menominee fought the British in the French and Indian Wars. When the French retreated from the New World, the Menominee only reluctantly switched allegiance to the British. Still, they had learned a lot about the Europeans since the days "the strangers" had first introduced them to guns, liquor, and bread. They had become political enough to know that the time for a shift in allegiances had arrived. Therefore they became loyal British allies, joining the British in their wars against the Americans. Instead of being a hospitable people in awe of white spirits, they changed into a people who recognized the political realities imposed by the superior force of European powers. They acted upon those realities in order to preserve their best interest. In this process they also learned how to adapt to the white man's ways while, at the same time, retaining their Menominee core.

The changes in the tribe's posture toward the late-French, and then the British, presence prepared the Menominee for their own course when they were faced with deciding what to do with their forest. What they decided was to protect their Menominee core by communally owning their land in spite of blandishments that could have led in another direction. Similar blandishments changed the histories of many American Indian tribes for the worse. At the same time, the tribe bowed to American economic power and joined the American economy in an effort to recover from the disasters visited upon them by the American treaties.

The Americans

The victory of the Americans over the British brought the Treaty Period of Menominee history. This period began inauspiciously enough. The United States established a trading post and Indian agency at La Baye (Green Bay) in 1815. The Treaty of St. Louis was signed shortly afterward. The treaty was simply a statement of peace and friendship between the tribe and the United States, but the changes presaged were ominous. The Treaty of St. Louis seized no Menominee land, but the mechanism represented by the treaty instrument would, by the Treaty of Lake Poygan, strip the Menominee of all their Wisconsin land holdings. During the same period, starting shortly after the Treaty of St. Louis, the fur trade, which the Menominee had integrated into their economy with the coming of the French, faltered, and then failed. The whole story of the establishment of private land ownership, the frenzy of logging the great Wisconsin forests, the insatiable demand by white American families for land and more land, the great building of a farm-dominated landscape, and the eventual founding and building of towns, villages, and cities all began to occur with an inexorable movement the Menominee were powerless to stop. In the midst of this the Menominee were decimated by diseases such as small pox and eventually starved into submission by forces put into motion the day the Americans replaced the British as ascendant in Menominee territory.

At the beginning of this process the Menominee were confused. First, hunts began to fail as habitat for many game animals was reduced by clear-cut logging and the creation of farm land. Then the fur trade, already made difficult because of overtrapping, began to fail. Fashion changed around the world, and though furs and pelts still sold, demand fell, reducing the trade's economic benefits on which the Menominee had become reliant. Then the American presence began to grow. What started during the French period as a vanguard grew from a trickle, to a stream, to a river of Europeans, all clamoring for Menominee land and the wealth represented by land and the great forests.

The Menominee response to this disaster was clever, having been forged during French and British periods of ascendancy. They used their culture, the need for consensus among all the chiefs representing the tribe's bands and clans before the tribe could commit itself to actions, complaining, wheedling, and the absence of leaders as weapons in the battles that followed. They did not go gentle into their decline. They understood American greed and avariciousness. The French had taught them well. The British had augmented the lesson. But in the end every stratagem was overcome.

In 1827, tired of the Menominee claiming that substantial chiefs were not present for treaty negotiations, thus invalidating the treaty in question, the Americans settled a long-running Menominee dispute concerning who was head chief by appointing Oshkosh to the post. The Menominee accepted this appointment because Oshkosh was the grandson of Old Carron, a powerful chief who had been successful in dealings with the Europeans. He was also a full-blood Menominee, not a man with mixed French and Menominee heritage. When Oshkosh was appointed, the American expectation was that the Treaty of Lake Poygan would settle the Menominee fate in Wisconsin. Fortunately, for the Menominee, Oshkosh was a great leader who signed the hated treaty but still managed to slip through the Lake Poygan snare to achieve the Treaty of the Wolf River and the Menominee's Wisconsin reservation.

As soon as the Treaty of the Wolf River was signed, however, and the Menominee had paddled birchbark canoes to their new home, carrying what few belongings they had with them, the Americans shifted political gears and ushered in a new policy effort that still echoes throughout Indian country today. The Americans fought ferociously to confine the Indians to reservations. After they succeeded, the reservations, along with the treaties that created them, became the "Indian problem." Throughout the treaty period the tribes had fought to retain whatever they could of the life they knew. What they retained cost the American government and American taxpayers money. The answer to "the Indian problem"? Assimilate them. In the words of Charles House in the *Green Bay Press Gazette* in 1959:

> For almost four centuries the American Indian has been in contact with the Euro-Americans but he has remained the Indian. A slow-moving but inexorable solution to the problems of a minority within a majority has been tried and tested a thousand times in history and it has yet to fail. The solution and perhaps the only complete one:

> Absorption. The minority must be absorbed into the majority. It is an automatic and fundamental solution to the problem of man's unkindness to man. (Peroff 1982, 9; House, 19 August 1959)

Make them a part of American society, and the problem will go away.

The story of assimilation efforts by the U.S. government has several aspects important to an explanation of Menominee sustainable development. Assimilation efforts and the Menominee response to those efforts have been the primary forces behind the forming of the institutions that have protected Menominee sustainable development policies. Assimilation efforts have also helped to create the complex Menominee culture with its strata of income and social groups. These strata, in turn, have created the intense political climate that is so difficult for non-Menominee to understand. The assimilation story is also the story of how the Menominee have persevered in their effort to maintain sustainable development as a center of Menominee life in the face of numerous efforts to lead the tribe down different paths. It also helps to explain modern sustainable development policies and practices, since those policies and practices were largely formed in reaction to the tribe's termination era.

The Forming of Early Contemporary Menominee Institutions

By 1908 and the passage of the LaFollette Act, the Menominee had become increasingly sophisticated in their American government dealings. They not only had learned how to use their old tactics of complaint, heel-dragging, and the need for consensus decision-making in their culture as tools against the Indian agents assigned to them, but they had also learned how to affect U.S. government decisions that they deemed contrary to their own sense of what was good for the tribe. The intelligent strategy was to send your best leaders, orators, negotiators to Washington, D.C., with or without Bureau of Indian Affairs' approval. Then get sympathetic senators and congressmen to support your viewpoint before Congress. Plead the tribe's case with all the tools and powers it could muster.

The primary institution evolved to implement Menominee political strategies was the Tribal Council, or, eventually, the General Council. The General Council is still a powerful Menominee institution, and it is, as it was during the late 1800s and early 1900s, an intensely democratic institution.

During the period when the first efforts were made to achieve Menominee harvest of the forest they owned, the tribe's leaders would call a Tribal Council meeting. All interested tribal members could attend. At the meeting the council would work through its agenda, making sure everyone who wanted to speak spoke. The meetings were traditionally all-day, or even two- or three-day, affairs. Then, after endless discussion, the council would reach consensus. This meant that a majority would become obvious and then opponents would bend to the majority's will, matching the tribe to the majority's opinion. If a leader was to be sent to the Indian agent, to Washington, or to Madison, Wisconsin's capital city, to present the tribe's will, that leader would be selected, and the business at hand would move forward.

As the 1800s ended and the tribe moved into the 1900s, this expression of tribal will ran increasingly counter to United States government policy. Not long after the Indian reservations were formed, religious and political forces in the U.S. started the long attempt to end the tribal way of life and decision-making. These were to be replaced with the American tenet that individual will in both social and economic life is more important than the group's will. According to this tenet, the best way to protect the group's long-term interests is to have each individual look out after his or her own interest. By allowing individuals to pursue their own interests, the interaction of individual wills, by the sorting through of endless options, eventually make the set of decisions most likely to strengthen the group's overall interest. This reasoning is in sharp contrast to the Menominee contention that only the tribe's will is important. Therefore individuals need to bend their own desires in order to implement tribal policies.

Early attempts to "civilize" the Menominee centered around efforts to turn tribal members into farmers with their own plot of land. These efforts were aided by efforts to end tribal culture. Some of these efforts were extreme. The speaking of Menominee language was banned in schools and strongly discouraged at church, in public functions, and at home. A system evolved that removed children from home for extended periods of time so they could attend boarding schools. Tribal members who embraced the Catholic Church and began the process of acculturation were favored within the reservation's economic system. The acculturated were pushed into roles of tribal leadership by the Indian agent and the Bureau of Indian Affairs.

Those who tried to acculturate received better schooling, housing, and jobs than those that held onto their culture, language, and traditional ways of living.

As time passed the Indian agent, working with what Peroff and the Spindlers described as the "elite" Menominee, began to form new institutions (Peroff 1982, 28; Spindler and Spindler 1971, 5). A tribal Court of Indian Offenses was established. The government appointed three prominent Menominee chiefs, Neopit, Ni'aqtawa'pomi, and "Chickeny" Ma'tshikine'u, as judges and gave them "jurisdiction over 'misdemeanors' committed by Indians, civil suits where Indians were the parties, and violation of the liquor regulations." The Court was also to enforce a new Indian Code which had, among its provisions, the following law:

> The sun-dance, the scalp-dance, the war-dance—and all other so-called feasts assimilating thereto; plural marriages; the practice of medicine man; the destruction or theft of property; the payment of or offer to pay . . . the friends or relatives of any Indian girl or woman (when taken as a wife), are declared to be Indian offenses. (Keesing 1987, 191)

A hospital was established at about the same time in order to help end "the practice of medicine man."

These new institutions, along with the sawmill, were designed to give the tribe a sense that they controlled their own affairs while continuing the dominant influence of the Bureau of Indian Affairs on Menominee life. The true relationship between the Menominee and the Bureau, however, is clearly outlined in the Ayer report in a transcripted interview between Ayer and A. S. Nicholson, Superintendent of the Menominee Indian Mills in 1913. Ayer was in Menominee researching Indian complaints because of complaints by several prominent Menominee against Nicholson and his management of the mill. The following quotation is taken from the response of Nicholson to an Ayer question asking him why those complaining about Nicholson were not currently working at the mill:

> The complaint on part of certain Indians against my management is not a fair one. The sentiment shown is purely manufactured, based on no reasonable shadow of substance. It is purely the effort of the dissatisfied and those whom I have been compelled to discipline in my capacity as administrator of law and regulation and teacher. It is simply the effort of the few unprincipled faction leaders. Seizing

upon every pretext, twisting everything that will suit their purpose, manufacturing false evidence, preaching waste of Indian money, not borne out by the record. Distorting every conceivable kind of thing to gain their ends in order to influence and poison the minds of their simpler fellow Indians. (Ayer 1914, 102)

It must be remembered in reviewing this passage that Nicholson was an employee of the U.S. government charged with managing assets held in trust for the owners of the assets, the Menominee people. His job was to secure profits and livelihood for the Menominee. Still, from his viewpoint he was the "administrator of law and regulation and teacher." He was the administrator who was compelled to administer discipline against those who made trouble for him "in order to influence and poison the minds of their simpler fellow Indians." This might be considered the raving of a man who had let power go to his head if it were not for the fact that Ayer eventually cleared Nicholson of all charges:

In fact, there is a tremendous effort here [in Menominee] to make bricks without straw; and it does seem to me that the present agitation and charges are a poor return for the efforts made to run a saw mill successfully and do it with nearly 50 per cent of labor that can come when they [the Menominee] please and go when they please. (Ayer 1914, 14)

Menominee complaints centered around a number of key sustainable development issues. Many Menominee were upset that the mill was hiring only white administrators and refusing to train Indians to take their place. As the elder Menominee Wyeskesit was reported as saying earlier in this chapter, the question was, "If this plant [the sawmill] was a paying proposition would I be poor; would I be hungry? The white people you see here, they are the people who have good things to eat." Concerns about whether the mill was being operated properly and if the forest was being managed in the right way were also expressed. This was the period when the white pine, hemlock, and hardwood forests were completely destroyed on 20,326 acres because Nicholson and his head forester, J. P. Kinney, abandoned the selective cutting rules of 1912. The forest mismanagement that started during this period eventually led to the 1951 settlement in the Court of Claims that put $8,500,000 into the Menominee bank account.

What this shows is that the Court of Indian Offenses and the hospital were formed with the specific purpose of moving the tribe down the path of assimilation. The sawmill and Menominee forest were totally under the control of the superintendent. The only real institution in the early reservation period was the General Council. In today's world Menominee Tribal Enterprises, the Menominee court system, tribal government, Menominee schools, College of the Menominee Nation, and the Menominee Clinic are important tribal institutions that have drawn inspiration from these earlier Menominee institutions, but the Menominee did not really own their own institutions until after the Menominee Restoration Act was passed in 1972 and the Menominee Tribal Legislature was seated in 1974. In 1913 and for decades afterward Menominee institutions were designed as part of the assimilation policy supported by a long succession of U.S. presidents and Congresses.

National Efforts at Assimilation and How They Affected the Menominee

"The most forceful direct attack on the continued existence of reservations" started in 1887 with the passage of the General Allotment, or Dawes, Act in 1887 (Peroff 1982, 12). Throughout the period when Menominee institutions were first forming, the Menominee were also fighting against becoming one of the tribes forced to accept the General Allotment Act's provisions. The Allotment Act authorized the President of the United States to end the ownership of land by tribes by allotting 160-acre plots to individual tribal members. If the tribe owned more land than was needed to accomplish this allotment, the remaining tribal lands were to be assigned to white settlers. The lands allotted were to be held in "trust" by the federal government for twenty-five years or until the individual was "competent" to hold clear title.

Following their old pattern of acceptance, then rejection of U.S. government actions, accompanied by complaints and efforts to distract federal authorities, the Menominee, at the urging of Indian Agent Ebenezer Stephens, Nicholson's superior, accepted the provisions of the Allotment Act by a tribal council vote. Then tribal leaders such as Neopit and Chickeney began to lobby for both allotment and tribal ownership of the tribe's timber resources. Tribal leaders also strongly emphasized in their interactions with the local Indian agent, congressmen, and senators in Washington the importance of logging to the tribe (Beck 1994, 190–95). By 1908 and the LaFollette Act's passage, reservation talk about the Allotment Act had died. The Menominee had escaped the first effort to assimilate them into non-Indian society.[2]

By 1934 and passage of the Indian Reorganization, or Wheeler-Howard, Act (United States Congress, 46 Stat. 984, 1934), the Menominee were well on their way to becoming the nation's wealthiest Indian tribe. Their sawmill was prospering, and their leaders were becoming increasingly sophisticated in reservation governance. The Indian Reorganization Act ended the allotment of Indian lands. By 1934 the policy was a widely recognized disaster, having led to unsatisfactory economic, health, and educational problems on all the reservations impacted. A political consensus had formed, recognizing the impossibility of assimilating Indians into American society through allotment (Peroff 1982, 13). The Reorganization Act also allowed Indians to organize under tribal constitutions, establish tribal corporations, and determine their own forms of tribal government. The Act's biggest weakness from the Menominee standpoint was that it put the Bureau of Indian Affairs in paternalistic control of tribal institutions. The act was still designed to assimilate Indians into mainstream society by allowing them to gradually master the tools of government while still continuing the government's paternal guidance.

This new tack toward assimilation was largely welcomed by the Menominee. As usual, however, their actions were not exactly in line with the new policy. A good example was the way the General Council and the Advisory Council, the body set up to advise the Agency Supeintendent on Menominee Affairs, pursued implementation of the Reorganization Act's incorporation provision. The tribe was not particularly interested in setting up corporations that could pursue individual lines of business. Rather, they proposed to incorporate the whole tribe, putting Menominee assets under Menominee control and going forth into the future as a people who were a business. To Agency Superintendent William R. Beyer, this plan was too communally based. He believed that communal life failed to lead to either industrial or financial progress. The Bureau of Indian Affairs as a whole believed the Menominee could not competently manage their own affairs (Beck 1994, 251). The Menominee idea of incorporation was contrary to the Reorganization Act's purpose, leading to a stronger communal, and therefore independent, presence rather than to assimilation.

Menominee agitation over the issue led to three important events. The first was the filing of the Court of Claims lawsuit that alleged government forest mismanagement during and after the time the Ayer report was absolving Nicholson of mismanagement. The second was the achievement of the Menominee right to veto the mill budget, and the third was the appointment of the first Menominee superintendent, Ralph Fredenberg.

The filing of the lawsuit was one of the great victories in the Menominee effort to maintain sustainability. It held the government accountable for actions not in keeping with the provisions of the LaFollette Act. It also allowed the Menominee to express the importance of maintaining their forest in an official setting and strengthened Menominee resolve to protect their forest forever.

Two of these accomplishments were, in retrospect, of less importance. The tribe tended to overuse its veto power over the mill's budget, causing resignation by a number of successive mill superintendents, probably harming, rather than strengthening, mill operations. Tragically, the tribe eventually rejected Fredenberg's leadership. When Fredenberg was appointed, power on the reservation did not rest in the agency superintendent's office, but in the mill superintendent's office. This meant that Fredenberg, even though he tried hard, could not effect the changes in mill operations and reservation life the majority of Menominee wanted. The result was that eventually the tribal membership grew dissatisfied with the progress being made on important tribal issues and began to agitate against Fredenberg. He was twice investigated and twice exonerated from charges stemming from this agitation. Eventually he resigned and was transferred to a similar position with a State of Washington tribe.

The tribe's victory in the Court of Claims suit in 1951 after fifteen years of litigation resulted in a $7,650,000 judgment for the tribe and brought the tribe's funds in deposit in the U.S. Treasury to $10,500,000. This victory, in turn, gave Senator Arthur V. Watkins, Chairman of the Subcommittee on Indian Affairs, the ammunition he needed to push forward the final assimilation policy faced by the Menominee. Menominee termination was discussed in the U.S. Congress as early as 1947 in testimony provided by the acting commissioner of the Bureau of Indian Affairs, but it did not happen until 17 June 1954 when President Dwight Eisenhower signed the Menominee Termination Act into law. This act simply said:

> That the purpose of this Act is to provide for the orderly termination of Federal supervision over the property and members of the Menominee Indian Tribe of Wisconsin. (United States Congress, PL 83-399, 1954)

What it represented was the greatest threat to Menominee sustainable development policies in the tribe's history. To the Menominee it meant even more, as Verna Fowler, president of College of the Menominee Nation, indicated in a recent essay:

> Conducted with paper and pen rather than force, this policy was not as bloody as many of the other policies (such as extermination, the relocation of Indian tribes, the slaughter of the buffalo, the reservation system, and the Dawes Act pursued by the Americans), but the misery, havoc, and cultural devastation was the same. Termination was designed to attain one goal: The eradication of Indian people. It almost succeeded with the Menominee people.
>
> With the flourish of a pen, the U.S. government decreed the end of the Menominee as a people. Gone was the federal Indian recognition and with it Indian rights and privileges guaranteed by treaties and the U.S. Constitution. Menominee communal land ownership disappeared. In its stead came loss of ancestral lands through sales and property taxes. The tribally supported hospital and clinic closed. Bureau of Indian Affairs education, social welfare, and protection programs ceased. No names were added to the tribal roll after 1954. To fill the vacuum created by loss came the sale of freshwater lake lots, property taxes, corporation stocks, Income Bonds, severe poverty, educational failure, and the migration of Menominee families to metropolitan areas. (Fowler 1995, v–vi)

The act's thirteen sections established a policy toward Indian people that eventually created all the havoc outlined by Fowler and more. Section three closed tribal rolls, officially denying tribal membership to any Menominee born after "midnight of the date of enactment" of the act. Section six allowed the tribe to

> select and train the services of qualified management specialists, including tax consultants, for the purpose of studying industrial programs on the Menominee Reservation and making such reports on recommendations, including appraisals of Menominee tribal property, as may be deemed necessary and desirable by the tribe in connection with the termination of Federal supervision. . . .

Section seven directed the tribe to "formulate and submit to the Secretary" a plan for future control of "tribal property and service functions now conducted under the supervision of the United States . . ." Section eight authorized the secretary to "transfer to the tribe, on December 31, 1958" the "title to all property, real and personal, held in trust by the United States for the tribe." Section nine generously waived the right of either the federal government or the State of Wisconsin to collect income tax due to the "distribution of assets made under the provisions of this Act," but declared that:

> Following any distribution of assets made under the provisions of this Act, such assets and any income derived therefrom in the hands of any individual, or any corporation or organization (set up by the Tribe) . . . shall be subject to the same taxes, State and Federal, as in the case of non-Indians. . . .

Section ten sealed the fate of the tribe as dictated by the U.S. Congress:

> When title to the property of the tribe has been transferred . . . the Secretary shall publish in the Federal Register an appropriate proclamation of that fact. Thereafter individual members of the tribe shall not be entitled to any of the services performed by the United States for Indians because of their status as Indians, all statutes of the United States which affect Indians because of their status as Indians shall no longer be applicable to members of the tribe, and the laws of the several States shall apply to the tribe and its members in the same manner as they apply to other citizens or persons within their jurisdiction. (United States Congress, PL 83-399, 1954)

As Fowler noted in describing the act's effects, "gone was the federal Indian recognition and with it Indian rights and privileges guaranteed by treaties and the U.S. Constitution." In its stead came economic, cultural, and social disaster and growth, along with other changes, that prepared the Menominee to establish today's reservation.

Two major institutions were created as the result of the Termination Act. Menominee Enterprises, Inc. was the most significant of these institutions. The Enterprises, as it was known on the reservation, assumed control over the forest, management of the forest, the sawmill, and Menominee efforts to prosper in the U.S. economy. It would have been the realization of the Menominee corporation dream pushed by Fredenberg and other Menominee during the allotment debate except for the fact that Wisconsin government, the Bureau of Indian Affairs, Congress, and some in the Menominee leadership did not believe Menominee had the skill to operate a major U.S. corporation. Experienced business people were therefore appointed to serve on the Enterprise Board of Directors, and huge blocks of stocks were put under the control of two trusts: The Voting Trust and the Menominee Assistance Trust. The First Wisconsin Trust Company of Milwaukee held the largest block of votes under this arrangement.

The second institution created was Menominee County, Wisconsin's seventy-second county, governed by a County Board elected by the voters within the new county's borders. The county's borders were conterminous with the former reservation's borders. County government was designed to replace the Bureau of Indian Affairs' reservation offices and provide the Menominee with democratic control over their governmental affairs.

Peroff describes both the "intended" impacts of the Termination Act and these new institutions and the "unintended" impacts. The major policy goal of Termination was to achieve Menominee assimilation into the mainstream of American life. According to Peroff, ten years after the act's passage, "Menominee elites held a distinctly optimistic view of the future." He quotes an upper-level manager of Menominee Enterprises, Inc. to make his point:

> In the eight and a half years since termination there has been a remarkable about-face, not only in the economy, but in attitude. We are proud to be standing on our own feet. In the next eight and a half years we'll advance at an even faster pace. (Boyd 1969; Peroff 1982, 164)

Peroff also points out that

> many Menominee assumed positions of responsibility and leadership in Menominee Enterprises, Inc. and county government. The resulting acquisition of administrative skills and political sophistication increased self-confidence and made interaction with non-Indians much less threatening than had been the case under the supervision of the tribe by the BIA. (Peroff 1982, 165)

New commercial ventures owned and operated by Menominee entrepreneurs were founded. Intermarriage between non-Indians and Menominee increased. Significant outmigration of Menominee from the reservation also began.

Most Menominee dispute that outmigration and the rise of a Menominee Enterprises, Inc. elite was an assimilation policy success. Tribal leaders like Glen Miller, a former tribal chairman who died tragically young, believed Menominee leadership skills developed at the Enterprises during these years led in the wrong direction since the Enterprises Board of Directors was dominated by non-Indian corporate interests that did not understand what the Menominee policy of sustainable community was all about. To Miller, Indian leadership comes from a "golden circle" of understanding that results from being a part of the Tribe's history, spirituality, and "Mother Earth." Indian leaders are always trying to build a bridge between the "golden circle of Indian life and the profit-thinking of the white man." To Miller, MEI's leaders ended up being part of the white man's world of profit. They had little understanding of the golden circle (Miller 1994).

Regardless of the perspective of Termination's successes, the unintended consequences of the policy are clear. Peroff lists them:

> economic instability, higher unemployment, decreased public services, aggravated racial discrimination, lower morale and personal self-esteem, deeper political alienation, and renewed factionalism, together with intensified opposition to tribal leadership. (Peroff 1982, 169).

The tribal court system was ended, the tribal hospital closed, and the past practice of hiring every Menominee willing to work at the sawmill was ended and replaced with an effort to increase operating efficiencies irrespective of the effect on the employment of tribal people. Menominee (once one of the wealthiest tribes in the country) became a pocket of poverty comparable to the pockets found on other reservations around the country. A culture of welfare dependency evolved.

The most significant unintended consequence of the policy, however, led to the policy of restoration, or self-determination, largely fashioned by the Menominee and then applied to national Indian policy. The trigger for this new policy was the struggle Menominee Enterprises faced in paying taxes on the corporation's forest lands. Under the agreement reached during the pre-Termination negotiations, the State of Wisconsin agreed to accept lower property tax rates than would normally have been imposed on forestland owned by Weyerhaeuser or some other large corporate landholder. The primary motivation for this was to preserve the Menominee Forest. Those involved with implementing termination understood that the full burden of property taxes would bankrupt Menominee Enterprises in a short period of time (*Menominee Tribe of Indians et al. v. the United States of America* 1984, 101a–102a).

Unfortunately, even this reduced tax burden was too much for the fledgling private business, with its history of putting workers before profit, to bear. The result was that the Enterprises and Menominee County faced bankruptcy shortly after termination. The solution reached by the Enterprises Board of Directors (which, again, was dominated by non-Menominee) and the Menominee County Board was to create a man-made lake, Legend Lake, and to then sell recreational lake lots to non-Indians along the lake's shores.

The Menominee by this point in time had fought ferociously to remain in Wisconsin, had avoided the Allotment Act, and had even managed to keep their forest intact in the face of the pine barons' power. Now they were poised to sell land their ancestors had preserved. They were ready, because of the white man's assimilation policy, to violate their grandfathers' sacred trust. The land was no longer held communally by the tribe for the tribe's benefit. It was now owned by a corporation controlled by white men, and the white men were going to begin to sell it off one small parcel at a time. One key tenet of Menominee sustainable development policy, that the ownership of land held by the tribe was inviolate, hung precariously in balance, ready to dissolve into the more powerful assimilation policy. The tribal storm that followed was predictable.

The storm, as Peroff points out, was caused by the founding of DRUMS (Determination of Rights and Unity for Menominee Shareholders):

> The most dramatic unintended impact of the act [the Termination Act] was the rise of a new organization. . . . The new political faction harassed, overthrew, and replaced the established Menominee governing elite. The primary goal of the new governing elite within DRUMS was repeal of the Termination Act. DRUMS instilled a new sense of pride and purpose in the Menominee people. Armed with a renewed self-confidence based on their Indian identity, the Menominee eventually were able to reverse the ill-conceived policy of termination. (Peroff 1982, 163–64)

In a decade-long battle that included daily protests at the office set up to sell recreation land around Legend Lake, legal maneuvering, and a dramatic march from Menominee County to Madison, Wisconsin's state capital, DRUMS fought tenaciously to end Termination. Ada Deer, the former head of the Bureau of Indian Affairs, led DRUMS's battle both in Washington, D.C. and on the reservation. After a little over a decade of DRUMS effort, "The

Menominee Restoration Act" (United States Congress, PL 93–197, 1976) became law.

This point in Menominee history was critical in terms of Menominee sustainable development for several reasons. The first is that the Menominee had successfully backed away from a policy that would in time have destroyed the sustainability of resources they had managed to maintain throughout the reservation period. The experience of tribes subject to the Allotment Act was now being shared by the Menominee. Land owned by individual Menominee and land controlled by the non-Indian Board of Directors of the Enterprises was slowly slipping from Menominee control. Non-Indian interests were threatening to become dominant on the reservation. Before land sales ended, 1,300 Legend Lake lots were sold, this in a county populated by fewer than 4,000 tribal members. These non-Indian interests, moreover, were more concerned with profit and efficiency than in long-term sustainability. In many ways the Legend Lake project was a financial success. By 1971 the Enterprises had earned over 1.7 million dollars in profit. Menominee County's tax base rose from an estimated 18 million dollars in 1965 to 32 million dollars (Peroff 1982, 150–51). Losing control over Menominee institutions, Menominee land, and any hope of Menominee control over their own affairs was what drove DRUMS so hard in their battle against the termination policy.

The victory of DRUMS also secured, for the first time, Menominee control over Menominee affairs. Sylvia Wilber, one of the DRUMS leaders, explained the Restoration Act to the Menominee:

> The heart of the legislation is to restore the Government's status as trustee of Menominee land and to restore to the Menominee the Federal services which were taken from us by termination. . . . The legislation itself provides for election; thus, there will be no repetition of the lack of consultation which marked the passage of the termination legislation. The legislation also contemplates a Menominee-BIA relationship which provides the Menominee with more autonomy than we had before termination, but which is still in the framework of relationships between the BIA and federally recognized Indians. (National Committee to Save the Menominee People and Forests 1972, 100)

Menominee land would be protected under the treaty trust relationship with the federal government. Federal services taken away by termination were to be restored, but both of these developments were to happen under a particular framework. The Menominee, who had sought autonomy since signing Treaty of the Wolf River, were finally to achieve autonomy. The relationship between the Bureau of Indian Affairs and the tribe was to be different than it had been.

This meant that the Menominee would control their institutions, forest, and other natural resources. The Restoration fight had been centered in the fight over control of the Enterprise which controlled the forest. DRUMS had achieved their victory by collecting Menominee-held proxy shares of Enterprise stock so they could counter the stock holdings by non-Indian interests. After they had won their victory, earning control of the mill and forest, they had no intention of going backwards in history. Menominee control was, at long last, a fact. From now on the Menominee would be able to preserve their forest without fighting the white man and his ideas of profitability and efficiency.

Restoration also confirmed to the Menominee the rightness of their culture and core beliefs. According to those at the heart of the DRUMS movement, writing just before passage of the Restoration Act:

> The history of DRUMS' growth is the story of the rebirth of the Menominee people at the individual, family, and tribal levels. For many tribal members who had been overawed and confused by the mechanisms of the Menominee termination, membership in DRUMS has spurred a growing comprehension of the nature and effects of that government policy. For other Menominee who had brooded or raged against the injustices of termination, DRUMS has provided an effective group forum through which they can take decisive action. For many Menominee, dependent upon MEI for employment and income, DRUMS has stimulated a courage to defy their corporate leadership. However, to all its members, DRUMS has demonstrated that mass, united tribal action is not only possible, but offers the tribe its only chance to end the oppression of termination. (National Committee to Save the Menominee People and Forests 1972, 67–68)

The Menominee had the power to restore what was theirs and to reverse the injustices that had plagued them for so long. They had the power to reaffirm what was important to the tribe in ways that could strengthen the tribe. The Menominee were no longer victims, but a people who knew how to take action. One of the chief actions they would take would be to reemphasize and strengthen sustainable development as tribal policy.

Menominee Politics and Institutions through the Termination Era

The story of Menominee politics, political character, and institutions through assimilation and termination is one of layers. The core Menominee political character was formed before Nicolet first found the tribe on Green Bay's shores. This character was based on balance, egalitarianism, cohesion, hospitality, and a form of democracy based upon merit rather than heredity. Disputes among tribal members were handled with moderation and forbearance. Tribal politics had a spiritual core emanating from a complex cosmology that taught the Menominee they came from the earth and were brothers to animal totems. Tribal life was centered in group opinions held about individuals and how individuals served the community's overall spiritual and practical needs. The earth's resources were communal property to be used only for the good of all tribal members.

When the French arrived, the Menominee began to adapt. The tribe's political and spiritual characteristics were kept intact, but tribal life was integrated into the world economy through the fur trade. The fur trade caused dependence upon bands that traveled far and wide in search of beaver and other animal pelts. Village became less important while family and band ties became more important. The Menominee also became less hospitable and more inclined to watch out for their own interests. The tribe was still a tribal society more interested in the group than the individual, but the concentration on smaller bands rather than on the larger phratries created divisions in the way tribal members saw the world. Each tribal member was still responsible for the well-being of the entire tribe; but for increasingly long periods of time, as the amount of fur-bearing animals in the Great Lakes region decreased because of over-trapping, the family and the band were the only tribe around.

The British taught the Menominee the importance of adaptation again when they came to dominate Menominee territory. By this point the Menominee had changed from being always generous, to a people who watched out for their own interest. They had become distrustful of Europeans, and they had learned to carefully keep their own counsel. The old spiritual ways remained, but an ever-increasing flow of white men were coming into Menominee territory. The Menominee were, for the first time, concerned about their land and their ancient way of life. Individuals became increasingly distrustful of non-

Menominee intentions and cooperated with increasing reluctance with the military power in place.

The Americans, when they came to dominate the Menominee, were more of a foe than either the French or British. They not only came to stay, but they brought with them treaties that rapidly took away Menominee land. The Menominee learned how to complain and protest with increasing bitterness. They developed a strategy that led them to first agree with whatever American proposition was on the table, but then dispute they had ever agreed to the terms on which the Americans claimed closure. At first, key chiefs were missing from those that signed the hated treaties. When the white man tired of that claim and appointed a chief, Chief Oshkosh, new delaying stratagems were developed. Menominee numbers were decimated by small pox and other European diseases. They allied themselves to the Americans during the Black Hawk War, but saw their alliance turn into bitter medicine as they lost all of their Wisconsin territory after being forced to sign the Treaty of Lake Poygan.

As the Menominee lost territory, the saw and ax tore through the North Woods and left farms in their wake. The Americans tried to make the Menominee farmers, but the effort, though it lasted late into the 1900s, failed again and again. The Menominee were not farmers. They were a forest people. As they died from disease and faced an increasingly uncertain future, they clung to old ways. They were a spiritual people with a rich culture and heritage. They were people of the earth. They lived in a world where spirits were everywhere. But at the same time they were increasingly becoming children of the Catholic Church.

In the Treaty of the Wolf River, won through a drum-beat of complaint, shrewdness, negotiation, and stubbornness, the Menominee finally gained what would become their permanent homeland, a reservation. Reservation life was hard and confining. The old freedom to hunt and gather was gone. In its place were payments due from the Indian agent that either never came or were lower than they were supposed to be. Starvation and disease was common. The people of the wild rice wondered if this was the last Menominee generation.

But again the tribe adapted. They developed a vision of preserving their forest forever while at the same time using it to help them survive in the new world. They complained and made trips to Washington and convinced the U.S. Congress to pass the LaFollette Act. They began to deposit money in the U.S. Treasury. They did not prosper in the white man's way, but they took hold of a hard life, maintained who they were, and developed a vision for their future.

Then assimilation policies tested them yet again. They were to become a part of the majority white society. The Catholic Church and the U.S. government joined together to eliminate Menominee culture, language, and Indian ways. A new form of political adaptation took hold. The Menominee split into different groups. The Catholic Menominee tended to prosper, at least in a relative sense, while traditional and Native American Church Menominee did less well. Still, the tribe remained Menominee. They were still people of the earth. Some had become Christians, but they were a spiritual people. They retained their land, forest, lakes, streams, and the Wolf River. They were the Menominee. They would sustainably develop themselves in their way, not the white man's way.

But the forces of assimilation were powerful. No matter what they tried, they were unable to seize control of the institutions that controlled them. They came up with different ideas and different tactics. They complained. They made trips to Washington. They made the Indian agents and the mill superintendents uncomfortable. They caused investigations like the Ayer investigation. They came up with the idea of incorporating themselves, of appointing a tribal member to a Bureau of Indian Affairs reservation office, of telling the white man they could not be farmers. But nothing worked.

If they had given up, they could have sold their forest to the highest bidder and been done with the battle. For a generation the tribal members would have improved their economic position in life, but that was not the Menominee way. Instead, they resisted the Allotment Act and tried to avoid the Termination Act. They maintained tribal control over their forest. But all of this was to no avail. Assimilation brought termination. They were denied the right to be a people, the Menominee. They were made into Americans. Their forest was turned over to a non-Indian Board of Directors who controlled the Enterprises. They created Menominee County in yet another effort to seize control of their own destiny. Their sustainable policies continued. They complained. They made trips to Washington, but it was all wrong. Termination split them into deeper factions than they had ever experienced before. At last, facing bankruptcy, the Enterprises decided to sell lake lots around a man-made lake non-Indian developers called Legend Lake. The Menominee leaders of the Enterprise and the developers called it the Lake of the Menominee Project.

Then came the storm of DRUMS. DRUMS claimed Legend Lake was wrong. They claimed Termination was wrong. They marched. They complained. They protested. Ada Deer became one of the most eloquent voices in all of Indian country. Then the miracle happened. The Menominee people won. Land sales stopped. The Enterprises was brought under the control of DRUMS and the Menominee people. There was self-determination, Indian control. The Restoration Act became law. At last they had control over their forest, lands, and resources. They became a tribe again. They turned their lands back to trust. They re-validated the treaties. They again became the Menominee.

The only problem was that the fight DRUMS had waged had been harsh. George Kenote, the Menominee who had led the way through Termination, had fought bitterly. He and his followers had believed the Menominee did not need a federal trust relationship. He said over and over again that the Menominee, by learning to stand on their own feet as American citizens, would, in the long run, learn how to prosper. He believed the Menominee could, in time, gain as a result of Termination, despite the losses Termination had occasioned.

The DRUMS position was that Termination had not given the Menominee control. It had simply substituted non-Menominee government control for non-Menominee corporate control, and they argued eloquently that such corporate control had resulted in considerably more disasters than had resulted from government control. They argued that the treaties signed by the U.S. government with the Menominee were sacred bonds and that the federal government's duty was to maintain a special relationship with the Menominee. They also argued that the government's old paternalistic ways, where government agents, rather than freely elected Menominee leaders, controlled the tribe's affairs and destiny were and always had been wrong. They fashioned a new "self-determination" policy where tribal restoration to federal status was put into place. The federal government was to provide its services through the granting of the same dollars granted to other tribes, but the Menominee were to control those dollars. They claimed that by putting the tribe in control, the tribe would make greater progress than it could ever make under government control.

Now the question was, What would they do with their control? They had always fought for sustainable development policies. They had retained a real forest in the midst of farmland and second growth forest. They were a divided people. George Kenote had fought hard for his viewpoint. He had been an honorable man with an honorable viewpoint. From DRUMS's perspective he had just been terribly, terribly wrong. The Menominee were restored, but a whirlpool of political differences existed on the reservation. Would they succeed? Would they continue what they had begun? Would self-determination and restoration bring the tribe a new beginning?

The Lobster Bucket and
How It Relates to Sustainability

At the beginning of Restoration the Menominee story was increasingly complex. On the one hand an enormous political victory had been achieved, but those who had been the victory's architects, Ada Deer, Sylvia Wilber, and Shirley Daly, were more controversial than their achievements. They were called "the three sisters" by some opponents in reference to the three witches at the beginning of Macbeth. At the same time, near unanimity on the reservation supported the rightness of what the three women leaders had achieved. Restoration held great promise for the Menominee. The question was, who would be in control of the institutions that implemented Restoration?

After its election in 1993, substantial problems were faced by the Restoration Committee, the committee established by the Menominee Restoration Act (United States Congress, PL 93-197, 1976) to implement the act's provisions. Menominee poverty was deep and unrelenting. The newly formed Menominee Tribal Enterprises, successor to Menominee Enterprises, Inc., was not profitable. The Menominee County Education Committee had just won the creation of a public school district that was to exist within Menominee County's borders even though both DRUMS's leadership and George Kenote had opposed the move. The task of implementing the Menominee Restoration Act was also large, especially given the fractionalization of the DRUMS membership and the rise of competing reservation political groups.

The Restoration Act's passage "signaled the official end of the policy of termination" in the United States (Peroff 1982, 227). It also opened the tribal rolls for those born after the rolls were closed by the

Termination Act in 1954, reestablished federal status for the Menominee Indian Tribe of Wisconsin, and transferred into trust tribal assets such as the forest and Menominee Tribal Enterprises. Individual Menominee were also allowed to turn private land back into trust if they chose to do so. Other provisions of the act authorized the tribe to write a constitution, form a government, and "ratified the Menominees' demands for direct control over their own political and economic affairs" (Peroff 1982, 228).

The Menominee did not endorse all the act's provisions. Non-Indians who had purchased property on Legend Lake were allowed to continue to own their land. They could also sell the land they owned whenever they wished. The State of Wisconsin also "retained" the authority to provide public services for non-Menominee residents and Menominee taxpayers of the county after the majority of the county's land had been put into trust. The Bureau of Indian Affairs had also persuaded Congress to not provide federal services to both reservation and non-reservation Menominee since such services were not provided to other tribes (Peroff 1982, 227–28).

One of the major reasons for the political turmoil that greeted Restoration was the feeling on the part of DRUMS leaders like Jim White that DRUMS had given up too much to get Restoration passed. From this viewpoint the land sales had been forced upon Menominee and were therefore illegitimate. The other accommodations made were considered to be just as bad. Many young people also felt that the Tribe had wandered too far from traditional Menominee culture. They believed this wandering had caused the troubles the tribe had experienced after the Long Knives had gained dominance over Menominee territory. These voices urged a return to the old culture and religion. They rejected accommodation and compromise. The Indian way was superior. The white man's way was wrong, especially for the Indian, and had to be opposed.

Dissident voices within and outside DRUMS led to political fracturing. Several competing groups struggled for power over the future Restoration had made possible. DRUMS's leaders believed their leadership had led to an enormous good for the tribe and that they should be rewarded for the years of voluntary, difficult effort. Those that had led Menominee County and Menominee Enterprises, Inc. were still around and still believed they had much to offer the tribe. Dissident DRUMS leaders, on the other hand, believed DRUMS's leadership had accommodated the white man too much in achieving Restoration and had to be opposed. New traditional leaders, mostly young men, began to plot out an entirely different vision of the future. They envisioned an Indian world made whole by again embracing the circle of the Great Spirit and all the aliveness and spirit making up the circle.

This fracturing of Menominee political life made reservation life even more complex than it had been during the time of the Spindlers' research. The Spindlers found one reservation elite built cooperatively by the Catholic Church and the Indian agents. Acculturation led to material success and comfort. At the Restoration era's beginning, multiple parties and voices were competing for influence. In addition the old structures of clan and band still existed. Part of the job as a leader was to support family, relatives, and friends. They could be trusted. They shared your heritage and beliefs. In addition the new traditionals were a force to be taken seriously. They were small in numbers, at least at the Restoration Era's beginning, but many were college educated and determined in their belief that their way was best for the tribe. They were among the most acculturated tribal members, but they rejected that acculturation and wanted to create a world where the tribe would again join the Great Spirit's circle.

Menominee institutions created either after Restoration or continued from the Termination or pre-Termination days only partially relieved the reservation's political tensions after the Restoration Committee, then the Menominee Tribal Legislature, came into being. The oldest institutions included Menominee Enterprises, Inc. that was renamed Menominee Tribal Enterprises. Menominee County continued as it had since the first days of Termination, and the tribal council tradition dating back into pre-European Menominee history continued. In addition to these institutions a new Menominee Indian School Board was created with the establishment of the Menominee Indian School District. A number of new tribal institutions were created by the *Constitution of the Menominee Indian Tribe of Wisconsin* (Menominee Indian Tribe of Wisconsin 1993) written by the Menominee Restoration Committee and ratified by referendum. Included were the Tribal Legislature, the Tribal Court System, the Tribal Law Enforcement Office, the Menominee Clinic, and Tribal Government. All of these institutions were put into place to serve fewer than 3,900 tribal members living on the reservation.

This is the social and political environment in which Menominee sustainable development policies matured.

An Extreme Version of Democracy:
Shirley Daly's Lobster Bucket

Shirley Daly, former DRUMS leader, former chairperson of the Menominee Restoration Committee, former vice chairperson of the tribe, current chairperson of the Board of Directors of Menominee Tribal Enterprise, loves to explain how Menominee, and, by extension, Indian politics works:

> Indians are like a bunch of lobsters stuck in a bucket. If one climbs to the top and starts to grab ahold of the bucket's rim, the other lobsters reach out and, quick, drag the lobster back to the bucket's bottom. Indians progress only when they all get together and make a miracle happen and lift up the entire bucket." (Daly 1994)

Menominee-style democracy is certainly not for the faint-hearted. Among non-Indians working on or living beside reservations, a common perception is that nothing positive could possibly come from Indian politics. Feelings are regularly treaded upon, whole families are attacked, often to the point of libel, and anyone brave enough to venture such storms must expect to fall from political favor sooner or later. Extreme language or positions seldom penalize tribal politicians, at least in the short run.

Several aspects of Menominee democracy make it appear, at least to non-Indians, to be extreme. The first is the level of political discourse. The fact that the three major post-Restoration leaders of DRUMS were characterized as "the three sisters" by their political opponents only starts the task of describing reservation political discourse. In a recent tribal election one candidate for the tribal legislature and who later won election was caught during the campaign hiring high school students to distribute literature accusing her political opponents of prostitution, theft, and other colorful behaviors. These charges were made by innuendo, but few on the reservation missed the basic message. These kinds of attacks do not make up the norm of Menominee political behavior, but they are also not unusual.

Nearly every election contains personal attacks designed to sting. These attacks are especially common during General Council meetings when the tribe's leadership has to stand up and take whatever is dished out. These attacks are often so emotionally draining that whoever is in power at that moment wonders out loud why they continue to serve their people. During one contentious General Council, Glen

Miller became so nervous he started smoking again halfway through even though he had managed to stay off cigarettes for a couple of years and was proud of his accomplishment. After that Council he commented he felt so burned by the rhetoric he wondered if he could go through another Council. He did, of course. Menominee leaders nearly always go through another Council as long as they can continue to get reelected.

Confrontations at General Council meetings are a second aspect of Menominee politics. Restraint is shown in such confrontations in the sense that one Menominee would hesitate to call another a prostitute in public, but such confrontations can be emotionally charged. Sometimes the emotional content of confrontation rises to such a pitch that the words spoken are difficult to follow. Such confrontations are almost always over disagreements about actions one party believes are against the tribal interest. An undertone of personal or family well-being can be present in such confrontations, but an admission that such gain can be had if the issue goes the speaker's way is rare.

A third aspect of modern Menominee politics is that an election is almost always in progress on the reservation. The Menominee County Board, Menominee Tribal Enterprises, Menominee County Sheriff, Menominee Tribal Sheriff, a number of county officials, the Menominee Indian School Board, and the Menominee Tribal Legislature are all elected. Multiple candidates usually run for most elected posts. Periods without an ongoing campaign are short, and almost all who want to run for office can convince themselves they have a chance. The attacking, confrontational style of campaigns means there are always dissatisfied voters who would like to see a change. One effect of the number of elections held is that every issue gets a full hearing. With all the candidates and officeholders running, no possible issue is ignored for long.

A fourth aspect is that elections normally have multiple meanings. If a tribal political actor can discover a way to gain benefit for himself or his family, he will often seize that benefit. This is not a universal rule of Menominee politics, but is often present. Serious policy differences also differentiate the candidates. These differences are seldom based on the rational choice policy model described by policy scholars like Anderson (Anderson 1994, 32–34). Reservation political actors sometimes act rationally to serve their self-interest, as would be predicted by the model, but the individual's preferences are never considered the "primary unit of analysis and theory." The tribe's interests are always assumed to be paramount, even while individual political actors are trying to serve their own, family's, friends', or allies' interests. This means that even self-motivated acts have to be justified by rhetoric that says the acts are for the tribal good. Other justifications are based upon basic rules of fairness or justice. Elite theory, which claims "that public policy is not determined by the demands and actions of the people . . . but rather by a ruling elite whose preferences are carried into effect by public officials and agencies" (Anderson 1994, 29–31) described most of the Menominee political situation up until the passage of the Restoration Act. On today's reservation none of the major theories that explain political system behavior hold.

As stated earlier, Deborah Stone's (1988) thesis, which claims that political paradox derived from basic community values and beliefs and the natural conflict between major values and beliefs, comes closer to describing the current reservation political system than other theories. Still, Stone's thesis does not explain the most important aspect of Menominee politics, consensus decision-making. As hard as the Menominee battle over issues and personalities, more often than not, after prolonged discussion and rhetoric, the tribe as a whole makes decisions: The Menominee Forest must be preserved forever, tribal sovereignty must be defended at all costs, and gaming revenue is good for the tribe. These are all decisions supported by a current consensus of the Menominee people. The road to such consensus is often brutal, and only the largest, most important issues reach the consensual stage; but, as with the old-time Menominee, consensus exists. It is still the hallmark of tribal decision-making. Eventually, on political issues, paradoxes are resolved, and a tribal rationality is imposed. Eventually the Menominee as a whole, through endless politics, rhetoric, attacks, and discussion, make an informed, rational decision about what course will best benefit the tribe.

Consensus does not always last forever, of course. The charged political climate with its players, agendas, and new ideas guarantees that every consensus will be tested and retested over time. On the other hand, once a consensus is in place, the consensual agreement becomes part of the tribe's decision-making process. The consensus is consulted in the context of any new decisions considered. If the consensus has a cultural/spiritual component, changes around even the consensual edges become difficult to achieve.

The last important aspect of Menominee democracy comes back to Shirley Daly's lobster bucket. In the end no leader can maintain power for long. Political actors like the deceased Hilary "Sparky" Waukau who have a large following and have the reputation for serving the tribe's interests surface again and again, sometimes changing from County Board or School Board or Tribal Council office to some other major office, but sooner or later all major political actors face defeat. Power will be stripped from them, they will find themselves at the bucket's bottom, and they have to start justifying their political come-back. Some fall by the wayside and never return. Others come back time and again.

In the end, as in pre-European days, all Menominee have a voice if they choose to use it, and all Menominee are equal. They may serve the tribe, hopefully with honor, but they cannot become the tribe. The tribe is not only larger than any individual or family, it is, unlike the American society's concept of government, larger than the Tribal Legislature or any other elective body. No political actor is ever allowed to believe that power is a personal right.

Competing Institutions

Each reservation institution has a particular purpose. This mirrors the old Menominee clan and sub-clan system where each clan had respon-sibilities to the tribe as a whole. The Menominee Indian School District is responsible for education services. Menominee County provides State of Wisconsin services and helps the Menominee deal with an increasingly restive Legend Lake population. Menominee Clinic takes care of tribal health needs. The Tribal Sheriff and tribal courts deal with minor criminal offenses and disputes that occur on reservation land, and the County Sheriff has jurisdiction over non-reservation county lands. The Tribal Legislature and government handle Bureau of Indian Affairs and other federal government programs. They also provide tribal funding for a number of programs ranging from the Maehnowseikiyah Drug and Alcohol Treatment Center to select College of the Menominee Nation programs. Menominee Tribal Enterprises manages the land and the forest while Menominee Gaming operates the gaming businesses.

This separation of institutional purposes does not mean the insti-tutions do not compete against each other, however. All institutions on the reservation compete in multiple ways with other institutions. The clearest institutional competition occurs between tribal and county governments. Both offer welfare programs and competing law enforcement agencies. Clear and unclear areas of responsibility some-

times result in either confusion or tension. Even the taxpayer-sup-
ported Menominee Indian School District has competition from the
Menominee Tribal School, the Delores Boyd Head Start Centers in
Keshena and Neopit, and, to an extent, the tribally controlled college,
College of the Menominee Nation.

The Tribal Legislature, based upon the tribe's constitution, is
the most powerful institution. The tribal chairperson is the reservation's most powerful political figure. But each institution, even NAES,
a satellite campus of an Indian-owned private college located in
Chicago, acts as a power center from which Menominee can earn
influence, work toward the betterment of family or friends, or follow
through on plans and ideas they believe can better the tribe or county.
All of these institutions, while acting as power centers, also cooperate
while fulfilling their basic missions. For example, arrests are often
made by consortia of law enforcement agencies working together
rather than by a single agency. The Federal Bureau of Investigation
(FBI) is responsible for investigating major crimes like selling drugs
and murder, and most arrests involve work done by both tribal and
county officers as well as by the FBI.

This mixture of competition, service, and cooperation results in a
highly politicized institutional culture. Individuals frequently move
from position to position within and between institutions, and institutional morale can rise or sink according to who is up or down within
the institution's power structure. A saying often repeated on the reservation is that the best way to keep your job is to do nothing. If you do
some good, according to the saying, you're bound to upset somebody
and get yourself fired.

Mostly, a sense exists on the reservation that no institution is stable. New policies, new ideas, and new efforts to implement policies
are constant. Political ferment is only surface deep, however. In the
end all policy changes are made incrementally. Shaped by consensus
values, long-term policies are extremely difficult to change.
Sustainable development policies are centered in the proposition that
the Menominee Forest must last forever as a resource that serves the
Menominee people and the circles of their brothers—the birds, fish,
and animals. A further consensus is that the most up-to-date scientific
and technological developments should be used in pursuit of that end.
This means that sustainable development policies in place are difficult
to change unless those changes clearly advance the science of managing the forest. This also allows technological and scientific advancements to be applied immediately without recourse to political debate.

Scientific and technological advancements become fair game politically only if one group or another perceives they harm the forest, the Menominee environment, or Menominee culture.

This surface of political change, underpinned by deep cultural and societal values that need a consensus if anything other than the most conservative of incremental changes is to occur, is difficult for most observers of Indian politics to understand. Most observers look at the ferment that surrounds every issue and micro-issue on the reservation and see the personal harshness of the political culture. They fail to see that what they perceive as brutal, raw politics is the working of a deeply conservative system that largely rejects policy changes. The Menominee bend to the will of American power, whether that power is political, economic, or cultural; but then they object, argue, debate, and work furiously to maintain the heart of who they are as a people. The majority may be Catholics, but they are Menominee Catholics, a tribal people who embrace their ancient heritage, aboriginal religion, and culture.

Consensus is built through a process of policy-winnowing and opinion-attrition. Every Menominee has an equal voice, whether a tribal leader or a sawmill day laborer. Each of the reservation's institutions, as the result of the tribal council tradition, must listen to the voices who wish to be heard. Earlier in this chapter Wyeskesit was very precise in his language when he said, "I want the views of this proposition placed before the tribe, and whatever action my people will take, then I will give my opinion." First, policy proposals are placed before the tribe. Policy proponents and opponents give their views, usually confronting each other with the passion of conviction and the rhetoric of tribal good, and then Wyeskesit and the other tribal members give their opinion.

Major policy battles migrate from one institution to another on the reservation and thus last a long time. A contemporary example of this institutional policy migration is the current debate over the proper uses of reservation gaming revenue. This debate started within the Menominee Gaming Board of Directors originally but was quickly joined by the Tribal Legislature. From there it migrated to the Menominee Tribal Enterprise Board, the Menominee Indian School District School Board, the College of the Menominee Nation Board, the Menominee County Board, and any number of committees. Tribal members thus have the opportunity to speak time and again. Different constituencies enter the debate from the viewpoint of different institutions.

This battle for the hearts and minds of tribal members will rage, often with bitterness and hard feelings, until, finally, the majority of tribal members begin to coalesce behind one policy. Then the rest of the tribe begin to "change" their minds. In the great debate over Termination and Restoration, even George Kenote, the Termination-era leader who believed fervently that by learning to live independently of government subsidy the Menominee would progress faster than they could otherwise, ended up testifying in favor of Restoration.

Menominee institutions are as deeply conservative as the political structure. The institutions compete ferociously, each one attempting to assert itself, but the competition is just another part of the overall political culture. Each institution is competing to preserve and enhance the tribe as a whole. Individuals benefit in the process, but all individuals are part of the lobster bucket. If they climb too high or become too prominent, they will find themselves back with the people, trying mightily to cause a miracle—the lifting of the bucket onto the next rung of the ladder of success.

The Lobster Bucket's Effect on Sustainable Development

The lobster bucket political institutional system's effect on sustainable development is twofold. First, an established policy such as sustaining the forest, based upon historical lessons and cultural and spiritual beliefs, ratified by a long-held tribal consensus, is a powerful force in reservation life. The only changes allowed are seen by the majority of the Menominee people to benefit the forest and the reservation's ecosystem. A belief often stated by the reservation's foresters is that the difference between the Menominee Forest and the Nicolet Forest is that Menominee decisions are based on silvicultural prescriptions derived from experience, science, and practice. The foresters serving the Nicolet have to modify silvicultural prescriptions in order to meet the Forest Service's multiple-use doctrines. This means the Forest Service is always trying to mediate between silviculture and the desires of cross-country skiers, hunters, hikers, snowmobilers, off-the-road vehicle enthusiasts, loggers, and a host of others (Pecore 1994, conversation). The duty of the Menominee foresters is to make sure the forest lasts forever. Therefore they are empowered to follow the best path possible to achieve that end.

The idea of development is a different matter altogether. The Menominee do not enjoy being poor. A comment in *Freedom With Reservation*, attacking the U.S. government's Termination policy, gives a clear idea about how Menominee feel about poverty:

> Clearly, with some $5 million of special aid, over $1 million in state aid, and a rapidly sinking economy combined with increasing health and education problems and a skyrocketing tuberculosis rate, termination has not been a success for the Menominees. It has been a rationally planned and officially blessed disaster of the U.S. Congress. (National Committee to Save the Menominee People and Forests 1972, 15)

The increased poverty brought by Termination was clearly considered disastrous for the tribe. The undesirability of poverty does not make reservation development the number one priority, however. Development project after development project is tried, but more often than not, such projects fail. Gordon Burr, a Stockbridge-Munsee tribal member who has worked for economic development on Wisconsin reservations most of his adult life, strongly believes they fail because Indians distrust the whole idea of development. He said in a recent conversation that

> white men don't understand it, but the Reservation Menominee have a lot. They have their woods, their lakes, their river, animals, fish, and all the freedom in the world. They aren't crammed together like the white man is in his cities. The Indians in the cities miss everything the Reservation has. An Indian can pick up his rifle and go hunting anytime he wants. The white man has to worry about private property, hunting season, and all kinds of rules and regulations.
>
> The Indians don't like being poor. They'd like to have a nice house and plenty of money to travel, but they don't trust the white man's economy, and economic development, that's from the white man, not the Indian. Therefore they worry over every proposed project and give it the once over and try to see how it's going to hurt them and their way of life. They want to protect what they have because so much has been lost. This means that the decisive action and good decisions necessary to making economic development projects work just doesn't happen. Instead the projects, even when they're good and could be successful, get worried to death, and when something actually does happen, people start taking a careful look to see who's getting the benefit over everybody else, or the business is run downhill by those who pushed it forward in the first

place simply because they aren't really interested in that particular
business after all. Economic development on Indian Reservations is
hard. (Burr 1995)

Not only does development often run into the political and institutional controversies that greet every new policy issue proposed, as
noted by Burr, but it also has to be a project that does not run counter
to consensual decisions already made. If the Menominee Tribal
Economic Development Department wanted to build a new potato
processing factory on the reservation and secured the funding to
finance it, the project would still not be built if it needed ten acres of
forested land for its construction. The forest comes first, not economic
development.

Successful economic development on the reservation has, more
often than not, had a cultural basis. The forest, as was shown in chapter 4, is part of Menominee culture. Therefore the sawmill and other
developments surrounding forest management, harvest, and protection have had a long-term success. Gambling is also an accepted
Menominee cultural practice. Hoffman described in some detail the
ancestral "aka'qsiwok" game "played for purposes of gambling,
either by two individuals or by two sets of individuals" (Hoffman
1896, 241). In the early 1900s, gambling establishments flourished on
the reservation until closed by federal authorities.

The one major exception to this rule has been the rise of the entrepreneurial class. Entrepreneurs have been able to flourish in spite of
the reservation's lobster bucket economy. Located on lands privatized
by Termination, several of these entrepreneurs have achieved a success that would be appreciated by the members of any chamber of
commerce in the nation. The rise of this class in Menominee society
breaks two old consensual holdings: The first being that all members
of the tribe must rise or fall as a group rather than as individuals, and
the second being that all land should be held for the tribal good by the
tribe. The rise of this class is clearly the result of the partial success of
Termination as an assimilation policy.

The major purpose of Menominee politics and institutions is to
slow down the pace of development unless it serves the tribal good
and protects the communal ownership of land and the preservation of
the Menominee Forest and ecosystems. The rules of such a political,
institutional system are not often kind to individuals, but they insure
that long-term Tribal interests are served sustainably.

6

Menominee Economics

Acculturation and Its Relationship to Income and Status: A Development Economy?

AS A FORCE that helps to maintain the lobster bucket and is integrally related to sustainability, the current Menominee economy is complex and, if the goal is to achieve economic equality with other American communities, unsatisfactory. This is true even though dramatic development progress has been made since the inception of reservation gaming enterprises. The economic complexity rises from the poverty the Menominee have faced, especially since the Termination period, the Menominee sustainable development ethic, and the fact that the reservation constitutes a society separate from mainstream American economy. The tribe's history and its relations with the different European and American societies that have dominated the tribe's economic affairs have also played a role. Also important has been the push/pull effect of the American economy on individual Menominee, creating what Spindler and Spindler described as an acculturative society. Successfully acculturated families occupy the top economic rungs while the less acculturated exhibit the range of problems associated with slums and other poor U.S. communities.

As has already been described, the Spindlers found five acculturation categories among the Menominee in 1971 (Spindler and Spindler 1971, 2–6). If the Spindlers were doing their study today, they would describe a more complex society. One example would be those individuals belonging to the native-oriented and transitional group described by the Spindlers. Although traditional people still practice the Big Drum religion, they have today been joined by new traditional people who are often highly educated and share economic well-being with the DRUMS and older leadership advanced by the Catholic

101

Church and the Indian Agency. Another example of a changing society is the growing entrepreneurial group. These people would be categorized by the Spindlers as being part of the reservation elite.
Therefore they, for the most part, would have to be subdivided from
the elite acculturated group. Still, their numbers are growing, and they
are becoming an increasingly important reservation social group.
Non-Indians living on Legend Lake would also have to be added to
the population description. Good questions can be raised about the
Spindler's categorizations, but they are useful in the sense that they
help to describe the social complexity of the reservation.

This acculturative class structure is made even more complex by
the number of ways contemporary Menominee earn income. On
the reservation three major Menominee controlled business enterprises: Menominee Tribal Enterprises, the Menominee Supermarket,
and Menominee Gaming; a number of small entrepreneurial businesses in retailing, services, and construction; and four governmental
agencies, tribal government, county government, the Menominee
Indian School District, and College of the Menominee Nation provide
substantial employment. Menominee also work off the reservation in
bordering towns such as Shawano. The majority of Tribal members
live and work off reservation in communities as close as Shawano and
as distant as Milwaukee, Chicago, Minneapolis, and Los Angeles.

The value, goal, and cultural differences held by each of these
groups, combined with the values that rise from different ways of
earning income, overlaid by the persistence and power of Menominee
culture, create tensions that are part of the complex political struggles
between individuals and institutions described earlier. These political
struggles, in turn, affect the tribe's economic prospects as well as the
incomes earned by different families. Further complicating this picture
are jobs held by non-Indians in Menominee-controlled enterprises and
the dollar multiplier effect created when Menominee spend the substantial income brought into the reservation in neighboring non-
Indian communities like Shawano and Gresham.

A Description of Income Sources for
the Reservation Population

The major income sources for individual Menominee living on the
reservation come from either government or tribal enterprises. The
total tribal government budget was $12,996,170 in 1994, down from

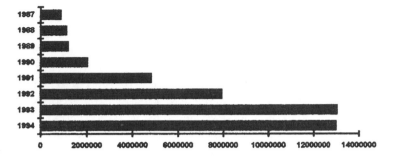

Figure 6-1. Menominee Indian Tribe of Wisconsin Budget, 1987–1984. (McDonough 1993, p. 78)

$13,039,808 in 1993. Figure 6-1 illustrates tribal government growth since 1987. Tribal government growth has primarily resulted from the infusion of gaming revenues into tribal coffers as figure 6-2, comparing budget numbers from 1988 and 1994, shows.

Gaming revenues were used in funding tribal programs for the first time in 1988. The total tribal budget for that year was $1,109,029. Gaming revenue equaled 37.2 percent of the tribal government budget. By fiscal year 1994 the tribal budget had increased to $12,996,170. Eighty-six percent of the budget was paid for by gaming revenue. Gaming revenue constitutes gaming enterprises profits minus any amounts spent on capital improvements. Tribal government, at least on the Menominee Reservation, treats profits as if they were a tax on the enterprises. In this case a huge proportion of enterprises' profits end up as funding for a variety of tribal programs ranging from college scholarships for tribal members to job training or social welfare programs.

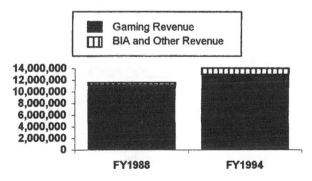

Figure 6-2. Tribal Budget: Government Income v. Revenue from Gaming, 1988 and 21994. (McDonough 1993)

By 1994 substantial gaming revenue had been used for construction. Completed projects by the time of the *1994 Annual Report* included Zoar Community Building, Woodland Youth Center, College of Menominee Nation Administration Building, Menominee Animal Shelter, College of the Menominee Nation Glen Miller Hall, expansion of Tribal Day Care, Maehnowseikiyah Treatment Center, Casino expansion, Menominee Hotel and Restaurant, Solid Waste Transfer Station, Food Distribution Program Warehouse and Office, extension of water and sewer lines, Keshena Senior Citizens' Center, and Tribal Utilities Building expansion. Projects fully funded and underway included Elderly Community Based Resident Facility, Middle Village Development Project, Tribal Office Building renovations, Menominee Logging Museum renovations, and Menominee Indian Head Start expansion. A number of additional projects had been undertaken with a combination of tribal and federal government grant funds (McDonough 1993, 78–79).

Menominee Tribal Enterprises, the second largest entity operating on the reservation, by contrast, showed sales less discounts in 1994 of $11,528,901 (Menominee Tribal Enterprises 1994, 17). Figure 6-3, constructed from MTE's Annual Reports for the years 1987–1994 shows the stability of this particular tribal enterprise.

Moderate income growth was realized from 1987 through 1994, from $8,265,813 to $11,528,901, but the real improvement has been in net profits. In 1987 MTE lost $1,010,500. In 1994 the net profit was $1,679,780. Net profits since 1984 are shown in figure 6-4.

The success of gaming enterprises and MTE during recent years has helped grow the reservation's entrepreneurial sector, primarily in the construction trades. Although no exact numbers are available, a number of growing construction trades firms are located on the reservation. Fewer than two dozen of these firms do business both on the

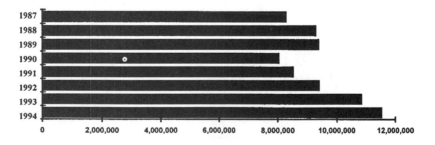

Figure 6-3. Menominee Tribal Enterprise Sales Volumes Less Discounts, 1987–1955. (Menominee Tribal Enterprise Annual Reports, 1987–1995)

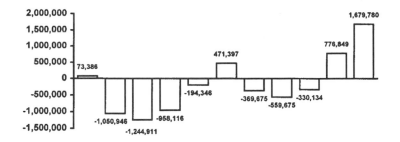

Figure 6-4. MTE Profit/Loss, 1984–1994. (Menominee Tribal Enterprises Annual Reports, 1984–1994)

reservation and in surrounding communities. The largest has a budget exceeding two million dollars annually. The smallest provides part or full-time employment for one or two members of Menominee families. They supplement retail entrepreneurs that have been operating since shortly after the Menominee agreed to the current reservation's boundaries. These businesses include gas stations, convenience grocery stores, bars, and women entrepreneurs who sell fry bread, Indian tacos, and other lunches to mill and tribal government workers. The number of these enterprises has also been expanding, but again, as with the construction industry, the contribution to the total economy is small.

The Scope and Depth of Menominee Poverty

Traditionally, Menominee County has had Wisconsin's highest county unemployment rate, and although some progress has been made (in May 1994, for instance, the unemployment rate on the reservation was 22 percent of the workforce compared to a May 1995 rate of 18.5 percent), because of the tribe's high birth rate, poverty still affects more than a fifth of the reservation population. In comparison, the May 1995 unemployment rate for Shawano County, immediately south of the reservation, was 4.8 percent. Menominee County, compared to other U.S. counties, also has one of the nation's highest county rates of children in poverty (Tempus 1995, 1).

The root causes of reservation poverty are difficult to eliminate even though the income earned by MTE, the gaming enterprises, the hotel and restaurant operation, and tribal and county governments has been increasing. Increased income has meant an increasing number of jobs, but many of these have been filled by non-Menominee even though tribal members need work and could fill some available jobs.

Included in poverty statistics are a number of educational indices that show Menominee students are not succeeding in school. Statistics from the Menominee Indian School District, the school system that serves most reservation students, show that as students pass through the grades, increasingly large numbers fall behind in reading and math skills. This difficulty with skill mastery in turn has led to a dropout rate of 9.5 percent of the entire student body per year. Only the Milwaukee school district in Wisconsin has a dropout rate close to the Menominee rate (Menominee Indian School District, 1992).

Other educational research has showed that developmental language delays affect 80 to 90 percent of all Menominee pre-school children (Pyatskowit and Madsen 1992, unpublished paper). Still other data has shown fetal alcohol syndrome (FAS) on Indian reservations varies nationally from 1 in 97 live births to 1 in 749. The prevalence of fetal alcohol effect (FAE) is more difficult to establish, but is thought to be roughly twice as prevalent as FAS (United States Department of Health and Human Services 1988). Menominee problems, according to Jerry Waukau, the director of the Menominee Tribal Clinic, are at least as severe as those outlined in federal studies (Waukau 1994, conversation).

A 1992 survey of Menominee youth, eleven and twelve years old, illustrates how prevalent alcohol and drug abuse is on the reservation. By the ninth grade only 18 percent of males and 7 percent of females claimed they did not drink alcohol. Forty percent of females and 61 percent of males claimed to drink alcohol recreationally. Overall, of the 350 young people surveyed, 57.6 percent reported the age of their first use to be 11–12. This compares to average first-use on a national level of 13.1 years (Youth Intervention Project Survey 1994).

Other statistics describing reservation poverty shows that 1,742 housing units serve a population of 3,890 persons. Of those housing units, 1,079 were occupied, according to the 1990 census. Fifty-five of the homes lacked complete plumbing facilities. Thirty homes lacked kitchen facilities. The mean wage and salary income for the 1,079 homes was $19,547 per year. A recent influx of Menominee from Milwaukee, Chicago, and other cities, have increased pressure on both housing stock and available jobs. The current population growth rate has been 15.3 percent per ten years. Recent rates, since 1990, have exceeded past growth rates (McDonough 1993).

Statistics can only touch on the problems faced by a substantial number of reservation families. Two scenarios constructed using facts taken from a number of different families creates a powerful picture of reservation poverty and its underlying causes:[1]

Scenario 1

The first family is headed by a single mother, age thirty-nine, who has a mild mental retardation. The father has been in and out of jail for fifteen years. The father has been arrested for domestic violence (striking the mother) and has allegedly struck the children. No case of child abuse has been proven, but there have been several referrals from schools and social workers. The family has two daughters and one son.

The oldest daughter is severely disabled. She has several different health problems, including mild retardation and several problems associated with fetal alcohol syndrome. This child also suffers seizures. This daughter is taking several different medications for her disabilities and requires twenty-four hour care; this includes toileting (incontinent bladder and bowel) and dressing and undressing. The other daughter and the son are healthy but are experiencing problems in the school system. These problems include disorderly behavior, truancy, poor grades, and absenteeism.

Neither father nor mother is working. The family is receiving AFDC and SSI. The mother knows her way around the welfare system and attempts to use every resource whether she qualifies not. She has committed welfare fraud in the past. Currently two hundred dollars a month is deducted from her monthly checks until she pays back approximately ten thousand dollars she received under false pretenses.

Scenario 2

The second family consists of a married couple with ten children. The mother has been hospitalized several times in the past few years for suicidal ideations. The mother wants to kill herself as the result of flashbacks about her childhood which she had repressed. She recalled incidents of incest during family-based therapy provided due to Juvenile Court referrals on the middle children. These referrals are for underage drinking, curfew violations, and runaway behavior. All of these children are also involved with gangs. One of the oldest

children is in jail. Another is living out of the home. All the
other children are missing significant amounts of time at
school.

The father does not live with the family, but resides in
Minneapolis. He admits to using drugs and allegedly sells
them. It is also alleged that the father is providing drugs to
his children. In addition, the father believes that in his culture
many mistresses are acceptable. The mother accepts this
belief without complaint.

This is the mother's second marriage. The first ended in
divorce. The mother is unemployed and is receiving AFDC
for herself and the children. She used to drink heavily but has
recently managed to avoid alcohol.

This mixture of drugs and alcohol, educational failure, values pre-
sumed to come from Menominee culture that do not come from the
culture, gang activity visited on the reservation from Chicago,
Milwaukee, and other city cultures, physical handicap, efforts to use
and misuse the welfare system, unemployment, despair, family histories filled with stories and incidents of incest and wife and child abuse,
the idea that families have moved to and from the reservation in an
effort to find a better life but failed to do so, and people out of control
of their own lives touches on the complex problems that make poverty
so difficult to resolve. These scenarios are neither unusual nor the
reservation norm, but those who have had trouble acculturating to the
respective strengths of Menominee and American cultures have ended
up on a bottom rung difficult for the tribe to put into the past. This history of problems and failure cycles from generation to generation
without any easing of either its pain or cost.

A Different Image: A Highly Successful,
Acculturated Family

The image left by the two scenarios should be contrasted with that of
the Alan Caldwell family. Alan Caldwell's mother, Letitia Caldwell,
was a successful real estate agent and served on the Shawano School
Board for many years. Alan, after graduation from high school, served
in the Army, graduated with a bachelor's degree from the University of
Wisconsin at Green Bay, and then became a professional educator,
holding positions as both a teacher and a consultant to school districts

for the Wisconsin Department of Public Instruction. He was also, for a short period of time, the director of Upward Bound for the University of Wisconsin at Stevens Point and the principal of the Menominee Tribal School in Neopit. He has a master's degree from the University of Wisconsin at Madison in school administration. His wife, Cathy, a Stockbridge-Munsee tribal member, is as educated and successful as Alan. She earned her master's degree in English from the University of Wisconsin at Madison, taught English for the University of Wisconsin at Stevens Point and the University of Wisconsin at Green Bay. She has written a children's book and has started working toward a Ph.D. She is also a published poet and has given poetry readings throughout the Midwest. The Caldwells are currently employed by the College of the Minominee Nation, developing a branch campus for the college on the Stockbridge-Munsee reservation. They have two children, both of whom have been successful in the schools they have attended. Alan also practices the Big Drum religion headquartered in Zoar, rejecting the Catholic religion of his youth. He is part of the new traditional category of Menominee discussed earlier in this chapter. The family is highly educated, articulate, goal oriented, and successful.

The Caldwells are still the exception on the reservation, but an increasing number of Menominee with either bachelors or advanced degrees are coming home and working in the reservation's schools, governmental agencies, and enterprises. This growing, affluent class of people are behind much of the current Menominee economic development success. They are often disunited politically, struggling for control over the reservation's government systems, but they are also, for the most part, determined to make the Tribal future more prosperous than the past has been. As can be seen by the description of the current Menominee economy, they seem, at least for the present, to be achieving a significant degree of success.

Summary of the Current Menominee Economy

Since 1987, the year when Menominee gaming enterprises became successful enough to contribute income to tribal government operations, the Menominee economy has been in a process of growth and diversification. Four primary sources of Menominee income existed prior to the start of gaming: income from jobs held off the reservation, jobs funded in one way or another by the federal government, a minuscule number of jobs provided by private reservation enterprises, and jobs provided by Menominee Tribal Enterprises. The unemployment rate exceeded 25 percent of the workforce. After 1987 the gaming enterprises began supplying an increasing number of jobs and income to both the reservation's tribal government and its people. The Aid for Families with Dependent Children (AFDC) federal welfare program caseload on the reservation has been decreasing (see figure 6-5). The State of Wisconsin welfare program, Relief for Needy Indian Persons (RNIP), has dramatically reduced its caseload (see figure 6-6).

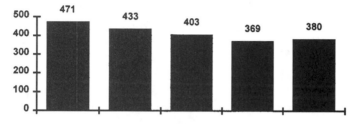

Figure 6-5. AFDC Caseload—Menominee Reservation, 1989–1993. (McDonough 1993)

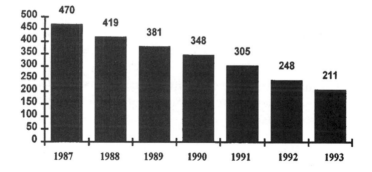

Figure 6-6. Average Menominee Reservation RNIP Caseload, 1987–1993. (McDonough 1993)

In addition, an entrepreneurial sector has been launched. The sector is still small but is growing. For the first time since the reservation's founding, income generated by individual businesspeople rather than by tribally owned and operated, or government sponsored, enterprises has become important.

Poverty, however, is still a powerful, complex force on the reservation. The social problems attendant on poverty have also been given a new dimension by some of the gaming industry's social problems. Gambling addiction has been added to drug and alcohol addiction as a powerful force in the Menominee community. These social and economic ills have been reduced by the growing Menominee economy, but they have not been eliminated, and no sign indicates they will be eliminated in either the near or distant future. Nor have the development successes seriously changed the pattern of acculturation and economic well-being documented by the Spindlers in 1971.

A growing educated population on the reservation, however, is not only achieving individual family success, but is also making a difference in the effort to achieve a better Tribal economic future. This group is divided politically, often holding different visions of the future. Much of the destabilization of Menominee politics comes from the different visions held by the different educated elites. Still, much of the success recorded by the reservation in recent years has been as the result of efforts by Menominee educated in universities and colleges from around the nation.

An Economic History

Key to understanding any current Menominee economics related to Menominee as a sustainable development society is a brief overview of Menominee economic history. This overview explains some of the economic choices made which have led to the current state of the reservation's economy. It also provides insights into why the current economy is, at least in some ways, part of the Menominee sustainable ethic.

Hoffman's description of Menominee economics prior to contact with European culture shows a people that lived off the abundance of the Great Lakes landscape. Game included black bear, deer, hare, porcupine, wildcat, and lynx. Beaver and otter were also present along with a variety of game birds, although grouse and ptarmigan were scarce. Sturgeon and trout were caught in great quantities and constituted an important food source. A number of other species of lake fish, such as whitefish, were caught from both Lake Michigan and inland lakes. The Menominee also harvested wild rice, blackberries, raspberries, blueberries, strawberries, maple syrup from maple trees, and a variety of roots and plants, often making food from what would today seem to be surprising sources such as the milkweed plant.

The animal harvest was accomplished primarily through use of bows and arrows and traps. Canoes, spears, nets, baskets, and a variety of tools crafted for specific purposes were used in fishing, maple harvesting, wild rice harvesting, and the collection of plant foodstuffs.

Manufacturing ranged from the making of bows and arrows to the building of longhouses and the making of baskets. Stone implements such as knives, needles, awls, and bowls were made as well as a variety of tools, carrying containers, and canoes made from trees and the bark of trees. Animal hides were used for clothing, blankets, bags, and the creation of art (Hoffman 1896, 272–94). As was described in chapter 3, land was used, not owned, and although individual items could be owned by an individual—for instance, a medicine bag was an item that individuals protected throughout their lives—a powerful sense of obligation to family and clan was more important than property. A strong obligation was owed to ancestors and the entire spirit world and especially the Great Spirit that had created the earth and all its richness.

Life patterns created by this type of economy combined sedentary and migratory aspects. For much of the year the Menominee occupied permanent villages. However, both hunting and trading with other groups such as the Winnebago or, later on, tribes as distant as the Huron, could take family groups or individual Menominee far afield. Religious events and social gatherings emptied villages and helped stir the pot of commerce between Menominee villages and the dwelling places of other tribes.

In many ways the manner in which the early Menominee lived matched the idyllic image of a people living lightly on the earth. They used their environment's resources in ways that did not seriously impact the environment. The number of Menominee and neighboring tribal people was small enough that their efforts to secure a living from the earth did not harm the earth. The few activities undertaken to modify the landscape—cutting down trees for longhouses or wigwams or creating a small pond by altering a stream's course—did not seriously harm the environment. The Menominee were in balance with nature.

But their lives were also difficult. The Spindlers tell a story from a later time that hints at just how difficult. The story starts when, after an absence from the reservation, the Spindlers travel into the deep woods to visit friends who were still practicing traditional Menominee ways:

Our friend did not seem quite as relaxed as we remembered her to be. Then the child on the bed began to cry restlessly. She got up slowly and walked hesitatingly to the bed to stroke her daughter's head. Then turned to us almost apologetically to say, "She's got some kind of illness. She's hot, and she hasn't kept anything down for two days." We did what we could, gave the child a little aspirin—and offered to take her in to the hospital to see the doctor. Then it dawned on us that last year there had been no baby and that the one swaddled in blankets looked virtually new-born. "Yes. She's new. She came three days ago." Then the story came out.

Our friend's husband had gone on one of his long trips to a neighboring reservation many miles away. In the meantime our friend had borne her fourth child during his absence and while everyone else was gone. She was isolated. She had no car and could not walk to the highway to get help. Her five year old daughter had become ill. There was no food left in the house. She said all of this quietly, with none of the hysteria or blaming that would have been normal for a woman in similar circumstances in our world. She merely stated the facts. (Spindler and Spindler 1971, 18–19)

As Father Jouvency wrote in 1710 in *Jesuit Relations*,

Whatever misfortunes may befall them [the Menominee] they never allow themselves to lose their calm composure of mind, in which they think that happiness especially consists. (Spindler and Spindler 1971, 19)

The early Menominee were in balance with nature, but they lived precariously within this balance, maintaining themselves and families with what we today would consider exceptional self-discipline, buttressed by a cosmology and rich religious system that helped them to understand the hardness and purposes of life and the self-discipline necessary to survival.

The reason this point needs to be made is that the early Menominee economic system cannot be contrasted with the system that developed through historical contacts with first the Europeans and then the Americans without an understanding that economics, religion, and culture were not separate in the way we think of those subjects today. Life was difficult, and the economic, religious, and cultural systems all arose from this one fact. You did not hunt or fish without involving the spiritual world. You apologized to the animal you had just killed in order to feed your family. When bad times came, you did not complain. You accepted your fate and prayed that the spiritual powers would come to your aid. In many ways economics, religion, and culture were the same thing. There was no economic man, religious man, or cultural man. There were human beings trying to survive in a world where winter nights could rage with temperatures from twenty to forty degrees below zero.

Writing about another aboriginal people, the Australian Abor-igine, Gary Snyder, an American poet and deep ecologist, explained this same idea in a way that not only helps clarify it, but also expands it:

> We were traveling by truck over dirt track west from Alice Springs in the company of a Pintubi elder named Jimmy Tjungurrayi. As we rolled along the dusty road, sitting back in the bed of a pickup, he began to speak very rapidly to me. He was talking about a mountain over there, telling me a story about some wallabies that came to that mountain in the dreamtime and got into some kind of mischief with some lizard girls. He had hardly finished that and he started in on another story about another hill over here and another story over there. I couldn't keep up. I realized after about half an hour of this that these were tales to be told while walking, and that I was experiencing a speeded-up version of what might be leisurely told over several days of foot travel. (Snyder 1990, 82)

In a world of pickup trucks, interstate highways, and jet airplanes, time has a different meaning.

Dr. Verna Fowler tells the story of how her mother would make her responsible for her grandmother when she was a little girl. The grandmother would think nothing of walking fifteen or sixteen miles a day, taking in the vibrancy of a spring meadow filled with wildflowers, in order to get her groceries (Fowler 1994, conversation). The economic system's unity with religious and cultural life had partially to do with this different sense of time and place. Memory was important. You had to remember where blackberries ripened in the fall if you wanted to eat blackberries. You needed to know when spawning sturgeon left Lake Winnebago and swam up the Wolf River. Without the stories that tied place, food source, and past events together, a mistake could be made during a harsh time that could be fatal to you, your family, or relation.

The French Economic System

In 1634, when the French came to Menominee country in the person of Jean Nicolet, the economic patterns Menominee lived by began to change. During pre-contact days, as Beck points out:

Menominee subsistence flowed in seasonal cycles of abundance and dearth; fortunately, when the rice or corn gave out during a hard winter, the sugar began to run in the maple trees, and the sturgeon and other fish began their run up the rivers that fed Lake Michigan and in the inland water systems. (Beck 1994, 290)

Food was harvested mostly for personal, family, or clan use. Trade, although it was important, especially from a social and cultural standpoint, was not central to the Menominee economy. All of that changed as a result of the fur trade. The Menominee continued to depend on traditional food, clothing, and housing sources well into the American period, but the French brought with them advanced manufactured goods such as tools, clothing, and weapons that the Menominee came to depend upon. They also brought alcohol, which the Menominee originally associated with the power of dreams and the spirit world,[2] and the twin ideas of credit and selling natural resources harvested from the environment to a larger economic system for profit.

This is not to say that the system of production and profit described by classical economists became the driving force in Menominee life. As is discussed later in this chapter, the Menominee still resisted American economic doctrines such as the transformative economy, but as contact with the French, English, and Americans imposed increasing demands on traditional Menominee life, the tribe slowly gained skill in surviving in the economic systems each culture imposed. Some sense of the traditional economic system still survives in Menominee interactions with non-Indian American society today. The Menominee Forest represents, in many ways, the aboriginal sense that religion, economics, and culture are unified. Neither the tribe nor individuals have the right to make an economic decision without first placing that decision within the powerful contexts of religion and culture. The Earth is your mother, and you do not purposely set out to wound your mother. But Menominee economic interactions with the three cultures that powerfully influenced the Menominee have also dramatically changed the old Menominee way of life. In some ways these changes have improved the tribe. On the other hand, important questions are being constantly asked in Menominee politics about just how much improvement has been derived from the American economic system.

The story of Nicolet's first encounter with the Menominee gives some sense of how immediately contact with the French began to change Menominee economic life. In the story, Nicolet introduced the Menominee to pots and pans, implements, firearms, and alcohol. The

French had goods the Menominee valued: "So now the Indian lived well, possessing these things" (Bloomfield 1928). The Menominee had talked to members of eastern tribes who had already secured these kinds of goods from either the French or English. Now they wanted guns, garments, knives, kettles, and hatchets for themselves. The French wanted beaver furs. They also wanted the Menominee, as they had wanted other tribes, to be dependent upon their largesse. The Menominee could provide the furs. The French could help outfit Menominee who were willing to trap by extending credit to individuals and Menominee families. The Menominee could become dependent upon French goods and credit. The system, out in the big woods of Menominee territory, seemed simple. The trade's complexity in terms of transportation, French and British politics, fashion, economic systems, and cultural and economic competition was not obvious. Still, all of these trade factors would affect the Menominee in a multitude of ways during the period between Nicolet's arrival at Green Bay and the signing of the Treaty of the Wolf River. A process was started that would lead to an economics compatible with the values, society, religion, and culture of a sustainable society.

The French credit system, introduced when Nicholas Perrot, a French trader, settled in the Green Bay area, is key to understanding both French era Menominee economics and later American development of a wardship Menominee society. Beck explains why the French credit system was key:

> The French credit system lasted beyond the fur trade into the twentieth century. The French extended credit for tools of the trade and for necessities such as food, cloth and cookware. At the end of the winter, the Menominee traded furs against their debts. . . . This extension of credit meant the Menominee could afford to spend more time pursuing the trade and less time on individual survival. This changed the nature of Menominee trading, which had a centuries-long history based on barter systems, and it also changed Menominee . . . hunting patterns [from the hunting of food] to hunting of fur-bearing animals on a commercial basis.
>   Two fur trade items stand out above all the others: alcohol and firearms. (Beck 1994, 291–92)

Firearms made hunting easier and more efficient. Less time had to be spent on survival. They also increased the effectiveness of Menominee warriors. By possessing and mastering firearms, the Menominee became allies rather than wards of first the French, then the British.

Firearms also led to a slow drifting-away from seasonal harvests and subsistence, tying the Menominee ever closer into the European economic system. Alcohol dependence had three main effects: First, it tied the Menominee closer to the economic system that provided the alcohol; second, it started the process of breaking down the iron disciplines needed for survival; and third, it used up resources needed for survival, threatening the lives of Menominee families and individuals. Alcohol-related deaths increased from poisoning, exposure, and murder. Alcohol also had profound effects upon religious and cultural practices (Beck 1994, 321).

Other serious fur trade effects included the liaisons and marriages between French traders and Menominee women. The Menominee were a particularly attractive people, and traders had little hesitation about fathering children with Menominee women. Menominee/ French marriages were common. European diseases also began to reduce Menominee numbers, complicating, making harder, and threatening Menominee lives. A time came when Menominee elders questioned their people's continued survival. A neighboring tribe, the Noquet, did not survive the European diseases (Keesing 1987, 8). The fur trade also brought Jesuits to Menominee territory. Their presence led to the conversion of many Menominee from their ancestral beliefs to Catholicism.

These were all effects caused by the economic system imposed by the French. The Menominee did not completely embrace the French system, but the attraction of firearms, alcohol, and other trade goods was powerful enough to get them to abandon ancestral subsistence patterns in return for indebtedness, loss of valuable tribal human resources, and dependency. Some tribal members resisted the French system. Few instances in Menominee history are recorded where some did not strongly resist change. However, most Menominee quickly absorbed the French economic system into tribal patterns and changed Menominee culture in order to absorb the new realities.

The new economic system did not always work well, however. Traders like Perrot and his partners often engaged in attempts to control more of the fur trade than they already controlled, and the French crown was always losing money on the trade. This conflict often affected the Menominee. An example was the late 1600s effort by the crown to establish a licensing system in order to siphon more of the trade's profits. By 1680 "as many as six hundred *coureurs de bois* lived in the woods as outlaws. Montreal's merchants happily outfitted these illegal traders . . . even the government officials were involved in the illegal trading" (Beck 1974, 302). For awhile the struggle between the crown and illegal traders made fur more expensive. Softened demand adversely affected the Menominee. Another instance occurred a few years later when the war threat in Europe and France's desire for greater control of the North American continent's interior led the government to embrace the outlawed *coureurs de bois*. Trade increased down the Mississippi to Louisiana. A glut on the European fur market resulted. Profits diminished throughout the French fur empire. By the 1720s this situation was reversed. The French government again imposed strict rules designed to regulate the trade and made it profitable again.

The Menominee did not fully understand how international politics and the laws of supply and demand impacted their lives in the woods, but it is clear that these were powerful forces. Even changes in European fashion taste flowed all the way to Menominee country, changing the price the Indian received for pelts and furs secured by trapping and hunting. The communal nature of tribal life changed, but did not disappear, under the new system's opportunities and drawbacks.

British Domination

By the time the British claimed Menominee territory in 1763, the fur trade was big business. During the American Revolutionary War the "greatest profits in the trade could be gained on the route to the west of Lake Superior" (Beck 1994, 326). The major difference between the French and British systems of running the trade existed in the size of firms that siphoned trade profits. French traders were either linked to the military or were individual entrepreneurs determined to make their fortunes. The British and Scottish-Canadian allies needed large capital sums to systemize the trade as its tentacles stretched ever farther west. This meant that merchants combined efforts and formed large concerns. By 1770 the number of beaver skins exported from Quebec to Great Britain had reached 102,920. Deer, muskrat, bear, lynx, cat, marten, fisher, fox, mink, otter, raccoon, and skunk pelts, and even swanskins were also sold as well as huge quantities of deer tallow and beaver glands for castor (Beck 1994, 327–28). This consolidation eventually resulted in a fur shortage around the Great Lakes, a falling off of the trade, and near monopolization of what was left of the

market. John Jacob Astor's American Fur Company eventually dominated the fur trade in the Upper Great Lakes region.

The Americans: Debacle

The British reign in Menominee country was short lived. The Treaty of
Ghent, signed in 1783, removed the British from Wisconsin and left the
Americans in charge. What had started as a Menominee desire for certain European manufactured goods and then a dependence on those
goods evolved into wardship. Treaty after treaty, forced by superior
American economic and military power, steadily reduced Menominee
territory until the Menominee owned no land in Wisconsin. The fur
trade waned first because of overharvesting and then because the
nineteenth-century logging frenzy destroyed much of the beaver's
habitat. The U.S. government began its attempt to make the
Menominee farmers. In 1836 the treaty negotiated at Cedar Point on
the Fox River provided, among other items, $500 for farming utensils,
cattle, implements, and two blacksmiths. Shops and materials for the
construction of farms or farm implements were included (Keesing
1987, 137). When the American hunger for Menominee land refused to
abate, the Menominee eventually found themselves government
wards on their reservation.

   Their economic life at this juncture was in a shambles. Ancient ways
of living had been completely destroyed, and, as the superintendent in
Green Bay wrote in 1854, the tribe was in a melancholy state:

> As far as I have been able to collect data in relation to the decrease of
> the number of Menominees, it appears that the decrease since twenty
> years is to be accounted for by the ravages of the small-pox in 1838,
> of the cholera in 1847 which latter was replaced by misery and star-
> vation—by men being killed in drunken rows, and by the fever,
> which, from time to time, commonly in the winter, has been raging
> among them, being clearly the consequence of want of provisions
> and other necessities; which was not alone their fault, as, since the
> first attempt was made to buy them out . . . until the present year
> have been such as not to encourage any improvements. (Keesing
> 1987, 144)

The tribe at this time lived primarily on treaty payments made to
families. Much of the tribe attempted to abandon their hunting and

gathering economy for the farming economy, but the new reservation's lands were poorly suited to agriculture, and past patterns of
subsistence often got in the way. When the time to tap the maple trees
for syrup came, it was time to tap the trees—even if it was also time to
plow fields. A few Menominee were successful farmers, but most were
more interested in timber than in corn and fat pigs.

Summary of the Meanings of Menominee's Economic History

Throughout their history the Menominee have retained parts of
their religion, culture, and historic patterns of thinking and changed
other parts, depending upon the cultural and economic environment to which they had to adapt. The Menominee are still acculturating, forming a culture neither purely American nor purely ancestral Menom-inee. This process has formed a complex society with
multiple layers of economic success and failure. It has also formed a
unique way of looking at the world and its economic powers that is
at the heart of the Menominee sustainable ethic. The Menominee
had much. They lost a lot. Therefore, they live not only for today
and the profit available from mining resources, but are interested in
shaping economic decisions to secure the future for their children's
children into forever's distance. They define the future's richness
partially through past tragedy. Gaming as an economic engine is
acceptable partly because its allows the Menominee to gain income
from the larger economy without compromising land-base and forest. All efforts to build a reservation farm, manufacturing, and
tourism economy have failed in spite of millions of dollars invested
by federal, state, and local governments. These activities have
brought the American economy too close to home. Activities rising
from these types of businesses might negatively impact either
Menominee culture or the land itself, and the land, the people, and
the spirit must be retained forever. Better poverty and hard times
than a loss of what makes the Menominee who they are and what
they will be.

That does not mean the Menominee reject change. They have
modified their behavior and outlook in dramatic ways since Nicolet's
coming. However, they take a conservative approach to change,
changing only under duress and only in ways that make sense to a

communal will expressed through political organizations, the tribal legislature, tribal courts, and General Councils. The economic system willed upon them by the French taught valuable lessons. Dealings with the British and Americans taught lessons as well. As was described in chapter 5, criticism is a powerful reservation political weapon that both shapes and inhibits change. Participating in the American economy today is a good as long as participation leads toward sustainability and Tribal continuance. Participating in the French economic system was a good, up to a point. Individual Menominee are encouraged to change more than the tribe itself changes in order to model what potential changes might mean to the tribe. The result is a system of checks and balances that impedes and pushes progress at the same time, always tending toward conservatism and careful, measured, incremental steps forward. The Menominee have been preserving what is theirs for a long time. Forever never ends.

The Menominee Concept of Land versus the Transformative Economy

This sense of forever is especially important in that it helps define the way the Menominee approach the uses of land. In many ways the Menominee sense of land ownership defines them more surely as a people and community than any other attribute. In contemporary America the conventional view is that property is part of the transformative economy. Property "is a discrete entity that can be made one's own by working it and transforming it into a human artifact" (Sax 1993, 162). As a human artifact, property then becomes part of the market system and can be bought, sold, improved upon, and used. It is an asset that can be maximized in value if used efficiently, or it can be devalued if its use is not efficient in market terms. Property values are set by market forces that work according to the economic laws of supply and demand. If a particular property is located in a desirable location a number of people would like to own, then the property's value increases as its value is competitively bid upward. If the property has little transformative value and is located in the middle of a wetland that cannot be developed because of environmental laws, then the property has fewer people desiring ownership and thus less value.

Herman Daly and James Cobb Jr., in their book *For the Common Good*, express this same idea in a different way. From their perspective modern economic thought has "reduced nature to land and land to space and matter." They go on to say that:

> from an early point [in the history of the discipline of economics] it shifted its attention from land to the rent of land, held rent to be price determined rather than price determining, and then successfully supported efforts to reduce its role in the economy. Thus land, which is the only element in economic theory pointing to the physical environment, does not function significantly within the contemporary discipline of economics. (Daly and Cobb 1989, 190)

The concept that individuals could own land, and that land (or the rent of land) was therefore a part of economic transaction, was not a part of early Menominee culture. The idea that land is part of a transformational economy would have been even more alien. A strongly developed sense of personal property allowed personal objects, or chieftancy status, to pass either matrilineally or patrilineally to descendants after death. Land was never passed from ancestor to descendant.

Hoffman's account of the tribe's founding when the light-colored bear came out of the ground after deciding to be an Indian shows just how powerful the Menominee land-sense was as a force in the tribe's early existence. The bear, who was the first Menominee, names other animals and birds as his brothers. These brothers became the heads of the first Menominee clans. Each brother also had a special relationship, as the bear did, with earth, land, water, or air. These special relationships between the tribe, its animal totems, and the elemental land, water, and air systems help to explain why the Menominee attachment to their land and its ecosystems is so powerful.

Beck suggests this story could have been metaphoric rather than literally believed by early Menominee (Beck 1994, 17), but whether such religious stories were literal recountings of what was believed or simply metaphors, the points derived from Menominee cosmology stories are still the same. The early Menominee felt a strong and powerful affinity to the earth's creatures, probably believing in the literal sense that their founding ancestors had been metamorphisized from Bear, Eagle, Beaver, Sturgeon, Elk, and other spirits. They hunted and fished the ancestors of their totems and clans, but were very careful to treat the animal and fish spirits that fed them with great respect, thanking the dead creature's spirit for its sacrifice on the hunter's or fisherman's behalf. They also thought of the earth that they, the animals, the birds, and the fish lived upon as grandmother, or mother, earth. Leonard Bloomfield, in his book *Menomini Texts*, explains, quoting one of his informants:

> The way of the Indian in the past, and his custom was this, that he did things even as the spirit powers gave it to him to do. And it was Me'napus, who was first given these herbs and roots, in all their various forms and as they taste. Then he in turn gave them on, that all the Indians might use them whenever they were ill, and grow well from their use of them. . . . Now this earth is the grandmother of us all; it is from her that these roots spring forth which this Indian is to us. (Bloomfield, 1928, p. 9; Keesing, 1987, p. 46)

How can you transform your grandmother into a human artifact that can be bought and sold? How can you even dream of selling parts of your mother or grandmother? How could you sell the dwelling place of animals and fish that were totems of you and your clan and thus bore an ancient kinship to you? The idea was impossible, and even after first contact with the Europeans, the concept of land-as-property did not become plain until after the American Revolution had ended, and the treaty period was well underway.

Beck tells the history of how the Menominee's misunderstanding of how the early American nation conceived of land-as-property led to formal treaty cessations. When tribal members finally understood what the white man was really after, they were amazed. Explaining how the Menominee originally came to sell nearly two million acres to tribes being relocated from New York to Wisconsin for five thousand dollars, Beck writes that:

> At first glance, judging by the astonishing amounts of lands ceded in the three treaties [to the United States government and the Oneida and Stockbridge, Munsee, and Brotherton Indian tribes], and the paltry sums received for them, it would seem that the Menominee who signed the treaties had little understanding of the American concept of land value. However, in the case of the first treaty, the tribe's leaders punished the signers, killing one of them. As to the next two treaties, the Menominee never thought they were selling the lands. As gracious hosts, they offered to share their lands, albeit reluctantly, as they had with the French and with the British. The gifts they received the Menominee probably viewed as the offerings of gracious guests, who were visiting them on their land. (Beck 1994, 361–62)

Land was to be used, not sold. But even its use was prescripted. The Bear and Eagle came out of the land and were part of the land. In the same way Indians were part of the land. You might fight your neighbor if he moved too close and threatened the way you were living—you might even drive your neighbor from land you saw as your own, but land was not personal property. Personal property could pass between the generations. The land was your grandmother. It was as integral to the Menominee community as the chiefs, sub-chiefs, or Masha' Ma'nido, the Good Mystery. In all its aspects—trees, animals, spirits, wild rice—the land was part of tribal life, sacred and good, there to be used by the Indians, the Menominee, to secure life, grandmother.

An increasing number of Menominee, the new traditionals, are attempting to revive the old beliefs as a contemporary religion, but most Menominee are practicing Catholics. They have learned the lesson about the value of land only too well. They have learned to defend ownership with a fierce ethic. Menominee who have moved away from the reservation to make a living in Milwaukee, Chicago, Minneapolis, or the nation's other cities often yearn to come home. With the rise of the gaming industry on the reservation, new jobs, combined with this yearning, have resulted in a substantial family in-migration. The in-migration has become so serious during the last few years the tribe has created Middle Village on purchased land formerly farmed and no longer forested. The tribe is determined to protect the forest lands against population increases. Land is sacrosanct. It must be protected. Although families are welcomed back to their homeland, they still have to live in a way that protects the forest and the tribe's other resources.

Are the Menominee Socialists?

During the 1950s and 1960s conservative commentators often accused the Menominee and other Indian tribes with a strong communal land ethic of being socialists, or perhaps even communists. The Soviet Union was at the height of its power during this period, and calling anyone a socialist or communist was serious business. The usual Indian rejoinder to such name-calling was that they could be neither socialists nor communists simply because their ways preceded by many centuries the lives of Karl Marx, Friedrich Engels, and the other economists, philosophers, and politicians who created such "isms."

The Indian viewpoint toward property was criticized long before Karl Marx ever read his first book on economics. John Locke once said that "there can be no injury when there is no property" (quoted in Chamberlin, 1975, 3). J. L. Chamberlin, in his book *The Harrowing of Eden*, quoted the chief judge of the Minnesota Supreme Court in the latter part of the nineteenth century who "put the case plainly":

> The Indians should be made to work for a living. Their land did not cost them anything. . . . The protection of all people in their just rights and privileges is axiomatic, but neither humanity nor justice demands that the Indians should be the pampered wards of our national government. If they will not work, let them starve. . . . The Caucasian has made marvelous strides in national greatness by reason of toil and thrift, and if the Indians can possibly save themselves from a fatal destiny, they must labor and save to deserve it. (Chamberlin 1975, 6–7)

As the description of the work of Joseph L. Sax, Herman E. Daly, and John Cobb Jr. indicated earlier in this chapter, much of the free market economy is based upon property ownership and the transformation of natural resources into wealth. The cost that will come due when natural resource stocks are depleted is never part of the free market equation. Therefore, the Indian attitude toward land is a frontal attack on free market economics. It rejects Locke's notion that land can be owned by individuals and emphasizes that land, and the natural resources found on or in the land, must be used for the people's benefit and not for the chosen few who have enough inheritance or earned capital to own land.

This story has another side, however. The Menominee have been willing to participate in the world's free marketplace through the sawmill and the development of the casino industry. A reservation entrepreneurial class has also been growing. The sawmill in particular has been a business with a social consciousness, often foregoing profits in order to keep greater numbers of Menominee employed. But, as can be seen from chapter 7, which describes Menominee Forest history, continuing efforts have attempted to improve productivity in the woods and the mill. The Menominee compete as hard as any corporate entity for the rational good of their businesses. Reservation entrepreneurs even engage in business in the time-honored American fashion by attempting to bootstrap themselves, their families, and the tribe to success.

In some ways the Menominee act more like a corporation than a
society. This analogy can be taken too far, but it is still worth making.
The Menominee want and expect their enterprises to be successful in
the marketplace. They are willing to reinvest and take steps toward
greater productivity as part of their drive toward success. The tribe's
real business, however, is not to benefit stockholders. Rather, its business is to benefit the Menominee people. From this perspective, the
employing of Menominee at the cost of short-term profit makes
rational sense simply because this is part of the Menominee people's
benefit package that is at the heart of business operations.

The answer to the question, Are the Menominee socialists? is No,
they are not socialists. Their overall viewpoint toward economics
comes from a different source than those of socialism/communism.
The Menominee believe they are part of the earth, as was established
in chapter 3. This imposes the responsibility to not only take care of
themselves and their families, but to also take care of the earth with
the same care a child would take care of an aging mother. Tribal land
cannot rightly be owned by an individual because (1) the produce of
the land (the natural resources) is properly dedicated to the benefit of
all the people and not just those who have been fortunate enough to
own land, and (2) the earth is too important to own. It belongs to itself
and not to mankind. As part of the earth's ecosystems, men have no
choice but to live from the earth's abundance, but they are also stew-
ards of that abundance. They have the responsibility to protect and
preserve that abundance forever. In *A Sand County Almanac* Aldo
Leopold argues that:

> The land ethic simply enlarges the boundaries of the community to
> include soils, water, plants, and animals, or collectively, the land . . .
> a land ethic changes the role of *Homo sapiens* from conqueror of the
> land-community to plain member and citizen of it. It implies respect
> for his fellow members, and also respect for the community as such.
> (Leopold 1966, 239–40)

He then points out that "obligations over and above self interest in
such rural community enterprises as the betterment of roads, schools,
churches, and baseball teams" are common throughout the country.
He finishes this thought by puzzling over why "land-use ethics are
still governed wholly by economic self-interest" (Leopold 1966, 245).

Recent scholars such as Robert V. Bartlett (1986) and John S.
Dryzek (1987) have introduced the idea of rational ecology, claiming

that in the same way economic choices can said to be rational when two or more individuals trade goods and services, choices that impact ecosystems should be ecologically rational. Dryzek defines rational ecology by saying that "the preservation and promotion of the integrity of the ecological and material underpinning of society—ecological rationality—should take priority over competing forms of reason in collective choices with an impact upon that integrity" (58–59). In other words, land-use ethics should be governed by the need to preserve the land community of which human beings are only a part. Economic, social, political, and legal rationalities must all fit inside this community's dynamics.

To a large degree these concepts are close to the systemic nature of Menominee sustainable development. The Menominee are intensely aware of the ecosystems that surround them and have developed choice mechanisms effective at preserving the communities of plants, soils, animals, water, and air that make up the ecosystems. Economic rationality is accepted, but it is bounded by an ecological rationality that has sometimes made the economic development part of the sustainable development equation difficult. Ecological rationality imposes costs that have limited the production efficiencies needed if full success within the economically rational free market model is to be achieved.

The Menominee, in short, are not after controlling the means of production so that the produce of production can be divided equally among the people, as would be true of a successful communistic economic system. They do not have any ideas about the redistribution of wealth. Rather, their primary motive is to preserve and protect the integrity of the ecological systems that provide their material livelihood. Land is communally owned to protect both the human inhabitants as well as all the other inhabitants of the Menominee land community. Both the Menominee and the other inhabitants of the Forest share in the communal ownership of Menominee land. The Menominee, in other words, have developed an ecologically rational economic system.

7

A History of the
Menominee Forest

The Birth and Practice of Menominee
Science and Technology

If HISTORY, CULTURE, spirituality, and tribal politics have been
driving forces behind the development and maintenance of reservation sustainable development policies, the practice of managing the
forest as a sustainable resource has had its own effects. The most
important of these has been the birth and practice of a science and
technology of forest sustainability. Chapter 4 contained a description
of some of the nonscientific, nontechnical topics that have been
included in the latest forest management plan. This chapter discusses
how the practice of sustaining the forest has changed from 1854 to the
present.

Paula Huff, the first director of the College of the Menominee
Nation's Menominee Sustainable Development Institute, often points
out that the reservation's database is far from adequate. She is a
wildlife biologist, and the scarcity of wildlife data concerns her. She is
especially concerned, as are Larry Waukau, Marshall Pecore, and others involved in the effort to sustain the forest and reservation environment while making it productive, that not enough data exists on either
riparian or forest biodiversity.

However, the existing historical and current data is extensive, and
even more importantly, the data is constantly reviewed as part of the
decision-making process used by foresters and scientists working on
the reservation and the business leaders of Menominee Tribal

Enterprises. A constant effort also attempts to broaden and deepen knowledge of the reservation, forest, environment, and ways of restoring damaged parts of the ecosystem. The goal is to make the Menominee ecosystem more functional from both productive and ecological standpoints.

The forest's history indicates, first of all, that neither science nor technology are necessarily enemies of either sustainable or development policies. Secondly, the history provides an insight into how ingrained Menominee determination, arising from their history, culture, and spirituality, is when the tribe is dealing with issues concerning environmental sustainability. Menominees have, again and again, protested, complained, and demonstrated against any effort, by tribal members or non-tribal interests, to harm the forest in the name of profits for either individuals or the tribe.

The forest's history also provides an understanding of legal issues the Menominee have faced in their effort to avoid assimilation into the larger American society. It sets the stage for a description of how the Menominee manage their resources in the contemporary world. These Menominee story elements need to be understood before a discussion about the applicability of Menominee sustainable development to the ecological, economic, cultural, and spiritual issues that face the modern world can occur.

The Periods of the Forest's History

Newman divides Menominee use of the forest's resources into five historical periods: Period I, 1854–1908; Period II, 1909–1925; Period III, 1926–1937; Period IV, 1938–1950; and Period V, 1951–1964 (Newman 1967, v). Period V really extended to the termination period's end in 1973, past the year in which Newman finished his thesis. Since the seating of the Menominee Restoration Committee, which wrote the contemporary Menominee Constitution, through the present, Menominee science and technology have become increasingly sophisticated. Tribal members have also achieved an increasingly important role in managing the tribe's various resources, leading to a flowering of the Menominee ethic, especially as it relates to the science and technology used to preserve and utilize the forest. For the most part, new technologies applied to the forest were the forces causing the end of one period and the beginning of another.

The First Period, 1854–908

The first period was characterized by the Menominee effort to gain the logging franchise for their own land. Once a reservation was established by the U.S. government, the government assumed the right to control reservation resources. As noted in chapter 3, the period from 1854 to 1908 was characterized by frenzied clearcutting of Wisconsin forests with the accompanying settlement by American citizens on farms cultivated after the loggers were finished. Early in the period, supervision on the reservation inadequately protected the forest from timber theft. Tens of thousands of board feet of primarily white pine located near the Wolf River were floated south for sale on the open market. Both Indians and non-Indians participated in this trade, stealing those trees with the highest value that were easiest to remove from the forest via the Wolf River.

The first timber-cutting authorized on a year-to-year basis by the Indian Commissioner occurred in 1854. The cut's amount varied during these years. The maximum harvest came in at three hundred thousand board feet in 1865. By 1867 a small mill capable of processing fifteen thousand board feet per day and powered by water was operating. J. P. Kinney in a "Report on Menominee Indian Forest" for Robert E. Mulroney, chief, Trial Section, Department of Justice claimed that one million board feet of timber was sold from 1871–1890, excluding local consumption (Kinney 1945; Newman 1967, 20). Most of this timber was stolen.

Pressure by the Menominee, and the further pressures exerted by constant non-Indian raids on the forest, led in 1890 to congressional action that allowed the cutting of timber for sale. The 1890 act was important in two respects. First, the timber barons hoped they could secure the Menominee forest logging franchise if the Indians were limited to what the barons believed was a small, insignificant harvest. A cut of twenty million board feet was set as the annual allowable cut. This cut was later incorporated into the more significant LaFollette Act of 1908. Apparently no sustained yield calculations were made to establish that this cut equaled the net growth of wood fiber in the forest on an annual basis, although later studies established that even though twenty million board feet was below net annual growth, it was still remarkably close to the amount of timber that could be sustainably removed given the time's science and technology.

The second reason for the 1890 act's significance was that it was the first law "regulating the cutting of timber on federally managed land in the United States." Although there "were no inventories or growth determinations . . . no organized forest protection program," and "silviculture was largely ignored, and sustained yield was not an objective" (Newman 1964, 20) this act protected the forest for a later time's more advanced concepts and started a process that led to Menominee sustained yield science, technology, and practice.

Two practices during this period that caused harm that still marks the forest today were high-grading silviculture and the leaving of significant amounts of slash in harvested areas. Both legal and illegal harvesters, for different reasons, participated in these practices. Although the 1890 act encouraged a clean cut in areas harvested, emphasizing pine and hemlock harvesting, many trees were left uncut. Uncut trees included many trees in good condition (Newman 1967, 23). This practice changed the species mix living in riparian systems on the reservation. Slash left behind in areas not cleanly cut resulted in fires that, combined with damage caused by fires occurring for much the same reasons during Period II, left some reservation areas with forest regeneration problems that have only lessened over time. They have not fully disappeared.

The Second Period, 1908–1925

The LaFollette Act of 1908 and Its Importance in the History of Menominee Sustainable Development

Period II was initiated by the LaFollette Act's passage in 1908. "Fighting Bob" LaFollette is a legendary Wisconsin political figure. He led Wisconsin's Progressive Movement and ran for the Republican presidential nomination under the Progressive banner in 1912. His autobiography expressed his attitude toward Indian affairs in 1908. When he first became a congressman, he had asked to be appointed to the Public Lands Committee, but the senior senator from Wisconsin, Philetus Sawyer, Wisconsin's Republican leader as well as one of the state's more colorful pine barons, decided otherwise. He influenced the House of Representative to tuck LaFollette away into the Committee on Indian Affairs instead. LaFollette's reaction?

> There was a reason for putting me on this committee [Indian affairs],
> and not upon Public Lands, which I did not appreciate until later. I
> had been too frank in expressing an interest in land grant forfeitures.
> It did not occur to Senator [Sawyer] that I might develop "foolishly
> sentimental" ideas against robbing Indian reservations of the pine
> timber in which they were very rich. (LaFollette 1960, 26)

After LaFollette was safely tucked into Indian Affairs, Sawyer then had one of his House allies, Congressman Guenther from Oshkosh, introduce a bill that allowed outside contractors, i.e., Sawyer's lumber company, to gain the franchise for harvesting Menominee timber. LaFollette was assigned to the subcommittee that was to mark up the bill. He had, according to his autobiography, been studying "diligently" the library of Indian books, treaties, and materials he had started accumulating as soon as he had been appointed to the Indian Affairs Committee. As soon as he saw Sawyer's bill he decided it was a license to steal Menominee timber. He went to see J. D. C. Atkins, then commissioner of Indian affairs, about the bill and was told by Atkins, "Mr. LaFollette, I think this is a little worst Indian bill I ever saw" (LaFollette 1968, 27). LaFollette asked for the commissioner's reaction in writing, then used the resulting letter to refuse to report the bill out of subcommittee even after Congressman Guenther objected.

This attitude culminated in the LaFollette Act of 1908 after LaFollette had been elected a senator in 1905. The act's key provision was that the Secretary of the Interior was authorized and directed:

> under such rules and regulations as he may prescribe in executing
> the intent and purposes of this Act, to cause to be cut and manufac-
> tured into lumber the dead and down timber, and such fully mature
> and ripened green timber as the forest service shall designate. . . .

on the reservation, provided that not more than twenty million board feet of timber "shall be cut in one year." This twenty million board feet was not to include footage harvested from "dead and down timber" (United States Congress 1908, 51).

The act also authorized building a reservation sawmill, the use of professional foresters from the U.S. Forest Service "in the protection of the forest," and the use of "Indians only" in contracts let for harvesting and employment in the new sawmill (United States Congress 1908, 51). As the 1908 *Congressional Record* makes plain, Menominee property rights and Menominee welfare were to be the primary considerations. The Menominee, not the pine barons, were to receive the "full value of this rich inheritance" (United States Senate 1908).

In his speech on the Senate floor in support of the 1908 Act, LaFollette said that:

> The forest is the natural home of these men. They are what is known as "Timber Indians." Their every instinct teaches them to seek a livelihood within the forest. The care, the preservation, of these forests should be the Indian's interest and his work. What the white man has in other places destroyed, the Indian should be taught here to preserve. This does not mean that the forest shall be permitted to remain in its wild state and contribute nothing to the industrial life of the community and add nothing of economic value to our country. It does mean that the harvest of the crop of forest products should be made in such a way that the forest will perpetuate itself; that it shall remain as a rich heritage to these people from which, through their own labor, they may derive their own support, and that, too, without ruthless destruction. Under the bill as proposed . . . these Indians shall be made a factor in our industrial life. In this way they will become self-reliant, learn to know the value of their heritage, and master the best methods for its preservation. (LaFollette 1908; Ourada 1979, 170)

Thus, the LaFollette Act and the 1890 act, both achieved after a succession of Menominee chiefs and leaders had visited and lobbied Congress, the White House, and the Bureau of Indian Affairs, established the foundation for the science and technology to follow. The Menominee would not repeat the white man's frenzied overuse of the land's resources. Rather, they would produce a "crop of forest products" in such a way that the Menominee Forest would "perpetuate" itself. They would become "self-reliant," knowing "the value of their heritage," by mastering "the best methods" for the forest's "preservation."

In the rapacious world of early-century politics these were radical concepts. As today's controversy in the Pacific Northwest continues to simmer over issues related to the spotted owl and preservation of remaining old growth forest ecosystems, these are still radical concepts. They encoded into law an idea of conservation that went beyond what Gifford Pinchot, the founder of the U.S. Forest Service, and Theodore Roosevelt, Pinchot's mentor, supported. Not only should the Menominee Forest be managed scientifically, but it should be managed in such a way as to preserve its attributes so that it would "perpetuate" itself. Preservation and perpetuation would even go further in order to help the Menominee know the "value of their heritage" as a tribal, woodland people.

This rhetoric from the beginnings of Menominee sustainable development blended elements that are part of the science, technology, and business of the Menominee Forest today. Menominee science is not dispassionate, nor is it always deductive. It is practiced with a clear historical and cultural vision in mind. The scientific method is used to conduct specific experiments and study in forest "demonstration" plots. Truth leads where it will through investigation, but in the end the science has a purpose: Forest, waters, soil, and other reservation resources must last forever. This ethic permeates decision-making about what studies and investigations are done and how they are conducted. It also affects the decision-making process that occurs during the course of any scientific investigation. If this ethic is not as fully scientific as it might be, then that is okay. It is a drum-beat rising from deep in the forest's, the people's, the spirit's heart.

During the second period of Menominee Forest use a number of developments occurred: The sawmill at the Village of Neopit was constructed in the Wolf River wilderness. The river-running of logs down the Wolf to the great lumber mills at Oshkosh and the other sawmills south of the reservation, with the almost miraculous dancing of Menominee lumberjacks on floating logs, ended. Fires made more intense by slash left behind by sloppy cutting practices and the great forest's blowdown in 1905 harmed the forest. And, finally, the railroad came to the reservation in order to help the new Neopit mill get hardwoods not amenable to river-running to market.

Both the blowdown of 1905 and the intense fires left major legacies. The blowdown, which blew down or severely damaged approximately forty million board feet, was mishandled by those in the Department of Interior responsible for managing Menominee affairs, and this mishandling led to the passage of the LaFollette Act, selective cutting, and the confirmed Menominee right to harvest their timber resources. The fires were destructive of the forest's long-term health. Newman quotes a letter to Senator LaFollette written by someone whose initials were J. F. B. that indicated an inspection of the Menominee pine forests in 1908 showed them "to be largely barren wastes without tree growth of any kind." He said that the Menominee pine forests had been mostly destroyed, leaving only the mixed hardwood, hemlock, and pine forests away from the reservation's waterways intact (Newman 1967, 27).

The First Major Technologies
That Changed the Forest

The building of the Neopit sawmill and the railroad were the first major technologies that changed the way the forest was managed, but these would not be the last technologies that played this role. For a while the sawmill and railroad centered reservation life in the Village of Neopit and allowed the forest's produce to be centrally processed. They also centralized marketing and forestry planning operations on the reservation itself, reducing the role Washington and Congress had played in both processes previously. The railroad's key value was the ability it gave the reservation to harvest hardwoods. Hardwoods cannot be floated down river to market. The railroad's coming ended the colorful river-running era and made hardwoods harvestable and valuable for the first time. The mill, located within the source of its supply, also increased the forest's value. With the building of the mill and railroad, the groundwork for Menominee management of the reservation was put into place, although the changeover to Menominee control did not take place until the election of a majority of DRUMS candidates to the Board of Directors of Menominee Enterprises, Inc. in the 1960s.

The Forest Service started the second period's other major innovation by marking trees to be cut down by logging crews working on the reservation. As long as lands were clear-cut, eventually to become farm land, identification of which trees to cut down and which to leave was not needed. Early in the second period, harvesting was supposed to clean up dead and down timber from the blowdown of 1905. This was not always done in practice in spite of efforts by the Bureau of Indian Affairs and the Forest Service to follow the LaFollette Act's directions. Some timber that was supposed to be harvested was not, ending up as unrecovered waste, and other green timber that was not to be harvested was processed at the sawmill. Later in the period the Forest Service and the Bureau started to work together to fulfill the LaFollette Act's intentions. Cooperation between the two federal agencies was not ideal, eventually leading to the partnership's breakup, leaving the Bureau, unfortunately, in charge, but initial efforts to create marking methodologies were developed by the Forest Service. This new science was called selection silviculture. The only experiments anyone had conducted in this new science in the United States, at the time, was at the U.S. Forest Service Lake States Region Experimental Station (Newman 1967, 31).

In 1909, E. E. Carter, a Forest Service assistant forester, wrote a letter to William Morris, one of the original timber markers, that outlined the challenge:

> The law requires the Forest Service merely to mark the timber which, in its opinion, should be removed for the good of the forest, and you must adhere to the letter and the spirit of this law, leaving all other matters, such as the selection of the areas to be cut, methods and seasons of logging and transportation of logs, disposal of brush and similar matters, to the discretion of the officer of the Indian Office in charge of logging operations on the Menominee.
>
> The aim which I believe must be the prevalent one, is to make an improvement cutting which will leave enough trees on the ground, sufficiently close together to protect them against windfall, to form in themselves a sufficient basis for a second cut, and at the same time to reseed the ground for future crops with the most valuable species on the reservation.
>
> The other method of marking applicable to the Menominee is to cut clear with the purpose of securing quickly a second growth sprout forest. (Newman 1967, 32)

Carter went on to state that marking had to be justifiable on both silvicultural and economic grounds. A tree was not to be marked if it met only one justification. He also said individual trees of poor vigor or in danger of windthrow should be marked in order to help keep the forest healthy (Newman 1967, 32).

The Ending of the Marking, Sustained-Yield, Experiment

The Forest Service's effort to mark trees was short-lived, however. In spite of the LaFollette Act's clear intentions, the Forest Service and Gifford Pinchot were gradually removed from their forest management role by the Bureau of Indian Affair's Menominee Indian office. This was partly the result of an exceptionally active fire season in 1910 that burned significant portions of three reservation townships. Mostly, the Forest Service fell afoul of the underlying profit motives driving BIA reservation operations.

The agreement between the Forest Service, under the Department of Agriculture's jurisdiction, and the Department of Interior, signed on 28 January 1908, two months before LaFollette Act passage, gave the Forest Service the responsibility for supervision of logging and timber sales, protection of all forests, and studying forests on Indian reservations to determine the best use of forest land. The Department of the Interior was to bear the expenses of labor, materials, and supplies necessary to the harvest of Indian forests and was to keep liquor away from Indians working in the forest and otherwise provide for the welfare of those Indians. All employed in the tasks undertaken for forest improvement and harvesting jobs were to be Forest Service employees (U.S. Forest Service and the Department of the Interior 1908).

From the very first days of implementation, the agreement did not work. The Forest Service was responsible for marking trees and maintaining the forest's long-term health, but the Department of Interior and the local Indian agent were responsible for all fiscal aspects of operations, including the sale of timber sawn into lumber. What this did was to give the Forest Service responsibility without control. They marked trees to be harvested, but the Indian agent decided on harvesting expenditures. Therefore the Indian agent controlled operations through the purse's power. H. H. Chapman, a professor from Yale hired by Menominee attorneys to investigate federal culpability in failing to follow the LaFollette Act in preparation for a U.S. Court of Claims case filed by the Menominee and eventually settled in 1951, said in an unpublished manuscript that by 1910 the Forest Service "had become heartily sick of the situation." They were looking for a way out of an untenable situation (Chapman 1994, 18). The 1910 fires gave the Forest Service the excuse it was seeking. The Department of Interior decided to clear-cut the burned-over areas and did not ask for marking in those areas. The Forest Service did not attempt to fulfill its responsibility even though clear-cutting was not confined to burned-over areas, but extended deep into forest surrounding the edges of where the fires had burned.

Chapman gave eight reasons the Department of Interior's logging superintendents persisted over the next several decades in clear-cutting the Menominee Forest in spite of federal law and a constant drum-beat of Menominee complaints:

1. They knew nothing and cared less about the practice of silviculture.

2. Their experience had been largely if not totally confined to clear cutting operations then prevalent in the region.

3. The installation of railroad logging appeared to them to require

clear-cutting in order to pay for the investment in the roads and spurs.

4. They were in sympathy with the so called system of "clear-cutting and planting," which resulted in clear-cutting and not planting.

5. They held that the Indian funds should not be expended on measures producing no sure returns for many decades.

6. They quickly became completely indifferent to the maintenance of fire protection and let the towers and phone lines cease to function by neglect.

7. Slash disposal, by burning during logging, to prevent subsequent slash fires, was, after an initial period, abandoned.

8. The need to show a profit in mill operations was interpreted by them as a necessity to remove every stick of timber that had a market value, resulting in "salvage" operations on previous cuttings.

(Chapman 1994, 10)

Chapman's conclusion about Indian Service's Forestry Department ascendancy in Menominee Forest management was disdainful.

The net result of the influence of these successive logging superintendents during the period following abandonment of selective cutting, by the rules of 1912, to its restoration in 1926 through J. P. Kinney's efforts and the forest department of the Indian Service, was the complete destruction of the white pine-hemlock-hardwood forests on 20,326 acres . . . for which destruction, specifically, the tribe were awarded the sum of $8,500,000 by the Court of Claims in 1950. (Chapman 1994, 10)

From 1911 through 1919 both Interior and Forestry sought to terminate their relationship by appealing to Congresses that consistently refused to amend the LaFollette Act as both parties wished, thereby relieving the Forest Service of responsibility. Using the 1912 rule, referred to by Chapman in the above passage, the Forest Service directed the Commissioner of Indian Affairs to clear-cut, in lieu of individual tree marking, "old growth timber in all forest types, including the pure hemlock type, the mixed hemlock, hardwoods, and pine, and the swamps." (Chapman 1994, 20). After publication of this rule, the Forest Service did not bother to send foresters to the reservation.

The Forest Service was not officially relieved of its abrogated responsibilities, however, until 2 March 1917 when an act (United States Congress, 39 Statute 991, 1917) was passed that allowed clear-cutting in the forest for agricultural purposes. A companion act (United States Congress, 40 Statute 561, 1918) was then passed on 25 May 1918 that authorized the Secretary of Interior to expend three hundred thousand dollars of tribal funds to help tribal members establish farms. This act referred to a nonexistent "Forest Service of the Indian Bureau." Chapman tells the rest of the story:

> On the basis of these references to the Forest Service of the Indian Bureau, the Commissioner of Indian Affairs apparently took the position that the 1908 Act requirement concerning the necessity for timber designations by the Forest Service of the Department of Agriculture had in some way been repealed, and no further requests were made to that agency for such designations for 1918 to 1928. (Chapman 1994, 25)

The Third Period, 1926–1937

What brought the Menominee Forest's second period to a close was the reservation's first roads. As Chapman indicated, railroad logging costs and operational centralization around the Neopit Mill became the justification for clear-cutting operations and LaFollette Act violations. Roads and the logging and hauling equipment made possible by roads allowed selective-cutting operations to resume. Roads also allowed the major advances in selective-cutting technology and science that dominated the forest's third use period.

Evolution from railroad logging to truck logging took place over a little more than a decade, but by 1936 the transition was complete. Government operations, centered in Neopit, were the slowest to change. Logging contractors, called jobbers in the logging industry, started using highways shortly after the first two reservation highways were built in the early 1920s.

The building of roads was important for three reasons. The first was that truck logging was more inexpensive than railroad logging. Roads cost dramatically less to build than railroad spurs into the woods. Thus, the logging operations could more easily access difficult-to-reach forest areas. The expansion of roads corresponded with the development of machinery with enough power and durability to haul and handle sawlogs. This meant logging equipment was not confined to areas close to expensive railroad spurs, but could be easily and quickly moved to different spots. This flexibility of movement, the third reason roads were important, connected harvesting technology to the forest in a way not possible before. Jobbers and government logging operations soon had a web of rough roads interlaced throughout the reservation, and the forest became, for the first time, manageable. It was no longer a wilderness with places inaccessible to those who wanted to mine the forest's resources.

Silviculture practices varied dramatically during the third period. During the period's first year an attempt at seed tree cutting was attempted. In burned-over areas loggers tried to salvage as many seed trees as possible while still cutting heavily into the timber stands. The idea was that the remaining "seed trees" would result in natural reproduction and an eventual reforesting of the area cut. That era's head forester marked 386 seed trees on 3,114 acres. This left about one seed tree for every 8 acres. The seed trees could not generate enough seedlings to compete successfully with brush, grass, and scrub oak (Newman 1967, 44). Thus, the experiment failed.

This experiment was followed by an experiment with diameter cutting limits. This was definitely not selection cutting forestry, but it was a step forward. It had been pushed earlier, after the 1910 fires, but in this particular experiment, conducted on 720 acres, a sincere effort was made to improve the forest harvest. The experiment was designed to make sure trees less than ten inches diameter at breast height and twelve inches at the stump would not be clear-cut, but would be left to regenerate the forest.

Designed by logging superintendent L. O. Grapp in 1926, this experiment marked the beginning of the selection cutting era that has preserved the Menominee Forest to the present day. Grapp's economic justification to the Bureau of Indian Affairs in Washington, D.C., for the experiment was as follows:

1. Hemlock was being logged and manufactured at an average loss of $1.03 per thousand board feet. Under the selection system, less hemlock would be taken, removing only the high risk trees and concentrating most of the cut on the hardwoods (He [Grapp] believed that hemlock would increase in value over the years).

2. Selection cutting would not take the low value small logs that

were so prevalent from clear-cutting operations. He considered
the fourteen-inch diameter as the optimal cut-off point; above that
size, the area necessary to get the allowable cut would be too great
from a cost standpoint. This, of course, was still in the railroad
logging period.

3. Under the clear-cutting system, artificial regeneration was necessary. He believed the high cost of planting could be eliminated by
dependence on natural regeneration which could be obtained
under the selection system.

(Newman 1967, 47)

Key to the process of moving toward selective cutting harvest
were experiments by the Goodman Lumber Company located in
Goodman, Wisconsin. At the time of Grapp's early experiments,
Goodman was also experimenting with selective cutting on a commercial basis. The results from the experiments seemed to be paying
off for Goodman. These results "gave reinforcement to Grapp's idea
and provided him with the necessary cooperation to launch" the selective cutting program (Newman 1967, 48).

During 1927–1928 Grapp attempted to remove only 70 percent of
the timber from harvest areas outside of where railroad logging was
continuing. The jobbers disliked these early efforts at selective cutting
and removed more than 70 percent. Other problems encountered were
residual damage to forest soils as the result of the widening use of
mechanized equipment, the great abundance of over-mature trees in
the forest, and the damage to trees left standing by logging operations.
Some species, such as hemlock and yellow birch, also suffered a high
mortality from exposure in cut-over areas after the forest's overall
density and canopy had been largely eliminated. In many cases the
forest tree mix changed after a 70 percent cut occurred. In hardwood
areas, for example, sugar maple and beech began to assume dominance over other hardwood species. In hemlock areas a conversion to
mixed hardwood-hemlock occurred since the hemlock did not regenerate.

These problems led to questions about the wisdom of a 70 percent
cut that led, in turn, to experiments where 50 percent of the forest volume in harvest areas was retained. These experiments had barely
begun, however, when major windstorms caused logging operations
to focus on salvaging dead, down, and damaged timber again. Salvage
operations continued until 1938. During this time efforts to continue

the 50 percent cut experiment were continued, but only on stands unaffected by the wind storms.

Menominee complaints about forest operations climaxed in the 1914 Ayer's Report that absolved all Menominee Indian Office personnel of any wrongdoing. Ayer, instead, accused complaining tribal members of drunkenness, laziness, greed, incompetency, and various forms of venality (Ayer 1914). Complaints, however, continued throughout the period. They lasted until a number of lawsuits accusing the federal government of failing to follow the LaFollette Act's provisions and the Court of Claims suit was filed in 1935. Reginald Oshkosh summed up the Menominee sense of their forest's history during a General Council meeting held on 9 August 1930:

> A lot of our old people now have passed away, those who have worked hard in order to save our properties, especially the timber. We can see now what the old people used to tell us, that some day the people would see how we have sacrificed many things in order to save our resources, i.e., the timber particularly. (Oshkosh 1930; Beck 1994, 530)

The Court of Claims lawsuit, although the final resolution of all the legal claims filed by the Menominee would not occur for fifteen years, forced the federal government to take actions to protect itself against future suits while defending indefensible silviculture practices that violated the LaFollette Act. As salvage operations wrapped up in 1937–1938, the fourth period began.

The Fourth Period: 1938–1950

In 1934 the reservation's last big fire occurred. After 1934 the web of reservation roads, advanced fire fighting techniques, and human vigilance on the reservation eliminated the age-old threat of severe fires. Elimination of this threat had one negative effect in that natural white pine regeneration requires fire for success. This would become, in time, an increasingly serious problem on the reservation since the Menominee highly value white pine canopies. The last great stands of old white pine on the reservation are aging and will be harvested during the next decade or so. But the elimination of major fires has been, overall, welcomed. The forest is too valuable an asset to allow major fires to decimate large sections even though fire was essential to the pre-American-era climax forest.

    The fourth period's beginning saw an effort to move harvest area cuts down from 50 percent of timber volume to 30 percent. A report by E. A. Sterling, commissioned by the tribe and released in 1934, found that if clear-cutting had been maintained at a pace of twenty million board feet per year, the forest's virgin timber would be depleted by 1968. Sterling also found that 10 percent of the forest had already been clear-cut and that 19,776 acres of clear-cut timber lands would not regenerate forest cover. He also stated that even if prompt and adequate reproduction efforts were undertaken to regenerate the forest in 1934, a fifty-year or longer gap would exist after 1968 before rotation age stands of timber would be ready for harvest (Newman 1967, 50–51). This study bolstered the Menominee position that the selective cut could be less. It also led the Indian Office foresters to support a lighter cut in practice as well as on paper—a far remove from actions taken during earlier periods.

    Richard Delaney, chief forester in 1938, revised the forest management plan resulting from Grapp's experiments with selective cutting. The forest had been subdivided into thirteen management blocks prior to 1937, the blocks established along topographic features, past cutting boundaries and blowdown areas. Delaney introduced the concept of a cutting cycle. The idea was that one block at a time would be harvested based on a ten-year cycle. The yearly harvest in the blocks would not add up to more than the twenty million board feet allowed by the LaFollette Act. The recommended maximum cut was also set at 30 percent of timber volume in any one area, allowing a lesser cut if conditions in certain areas needed more trees for reproduction or overall forest health.

    Marking rules were also more concerned with long-term forest health than had been true previously:

1. Dead hemlock will be marked, but if merchantable will be tallied.

2. Trees must have at least one merchantable log to be marked.

3. High risk trees should be marked, if desired volume to be cut can be achieved in dead or defective trees, do not cut healthy trees.

4. Minimum size limit guides for healthy trees (in inches):

Beech	18
Birch	20
Maple, Basswood	24
Pine	36

5. Mark hemlock as lightly as possible, marking only high risk and stunted individuals. In mixed stands, attempt to leave hardwoods.

6. As a rough guide, 7–10 trees per acre should be cut.

These marking rules noted that different species demanded different treatment and should not be treated the same. They also, for the first time, expressed concern about healthy trees. At this point the idea at the heart of the LaFollette Act and Menominee desires, that the forest could be maintained forever, had a place in a forest management plan. This period's foresters were still handicapped by a lack of scientific data on the forest. Many of their decisions reflected that handicap, but the true beginning of a science of sustainable forestry dates from these years as foresters battled mortality from windstorms, insect infestations, small fires, and other problems while in the midst of what must have seemed to be a continuous round of salvage operations.

The fourth period also saw an increased emphasis on planting efforts. In 1942 a token project was launched that involved Menominee men, women, and children in a voluntary effort to plant trees. From 1942 to 1950 annual planting efforts put seedlings into areas of eighty-two acres and less, though the effort dwindled to the point where only seven acres were planted in 1950 (Newman 1967, 62). Planting success during these years led to a full-scale planting program that started in 1952.

By 1938 nearly all the forests in the Great Lake States had been cut over twice. Pines had been harvested in the late 1800s, then, from 1900 through 1930, hardwoods had been harvested. Forest resources left in the region contained mostly aspen and second growth hardwoods and pine. By 1938 most of the big sawmills had shut down, taking with them the jobs they had created. Only three big mills were left in Wisconsin, one of which was located at Neopit. Mistakes, recounted in this historical survey, had been made in the Menominee Forest. Menominee had complained about the mistakes for decades and had eventually, out of frustration, sued the federal government with the objective of preserving the resource they owned, but did not control. Still, by the Fourth Period's end, the Menominee Forest was perhaps the most impressive forest of its type left in the United States. Its canopies, weaving pine branches together to shade the deep blue of Wisconsin summer skies, its great stands of maple, a fire of red, orange, and yellow leaves in the fall, and its hemlock, basswood, black walnut, cedar, oak, birch, and other species stood as an impressive forest, deep in both beauty and sense of pristine wilderness, even though no pristine wilderness truly existed.

The period's last significant innovation was the beginning of an effort to increase knowledge of the forest. The most significant cruise study completed before 1946 was finished in 1914. In 1941 a line-plot cruise inventory was started where twenty one-fifth-acre plots were tallied per forty acres for a 10 percent cruise. The objective was to study an area with the cruise inventory each year comparable to the area cut during that year (Newman 1967, 64). Then, in 1946, a Continuous Forest Inventory (CFI) was begun. This effort established 387 permanent plots representative of 132,000 acres of old-growth hardwood and white pine timber. Old IBM sensing cards were used to compile data. The timber types for each plot were established; then the cruise was implemented. The first measurements were made in 1954.

This particular innovation started the build-up of forest data that continues today. It also marked the beginning of an effort to understand the timber resource as a whole rather than to treat it as separate historical plots. For the first time the forest could be mapped in a way that accurately described the complex intermixture of reservation soil, water, life, and timber resources.

The Fifth Period: 1951–1973

The fifth period was marked by reservation political and social upheavals that in turn strongly affected forest management. The period started hopefully enough. The 1938 management plan was completed two years after the ten-year cycle, and the experience gained from cutting and logging during the cycle had led to a number of newly evolved objectives:

> 1) a lighter cut in the hardwood and pine types allowing more complete coverage of the forest in fifteen years, the time period chosen for the cutting cycle; 2) a 5-year cutting cycle in the hemlock type to most efficiently salvage mortality which occurred as an aftermath to cutting operations in the type; and 3) the inclusion of the by-products cut in the cutting budget in a more organized fashion. Along with these aims, a stepped-up planting program, an improved protection program, and the introduction of timber stand improvement prescriptions were vital components of the plan. (Newman 1957, 77)

These objectives continued the path set earlier that saw the forest as a living organism whose parts had to be treated differently, depending upon tree species and other factors. They also introduced, albeit in an unsophisticated way, the idea that the forest's problems not only had to be identified and addressed, but had to be prescripted. A plan, properly implemented, could ameliorate problems in a way that restored the forest's long-term health and productivity. Increasingly, foresters were seeing and treating the forest as a whole rather than concentrating on economics realized from harvesting board feet. The forest left alone to deal with the stresses of human habitation was not forever. Only good forest management could keep the forest alive for the long-term future.

Then, on 13 July 1951 the U.S. Court of Appeals awarded the Menominee $7,650,000 as a result of the court case filed in 1935. "The crux of the case," as H. H. Chapman described it,

> was the words in the LaFollette Act of 1908, authorizing the Secretary of the Interior to cut and manufacture into lumber only such *"fully matured and ripened green timber as the Forestry Service shall designate"* for *"the preservation of the forests"* [Chapman's emphasis] on the Menominee Indian Reservation. (Chapman 1994, 38)

The Court found that from 1910 to 1928 the federal government, in clear-cutting twenty thousand acres, had violated the LaFollette Act:

1. by cutting green timber that was not fully mature and ripened,

2. by cutting green timber that was not properly designated by the Forest Service of the Department of Agriculture which was the Forest Service referred to by the 1908 Act, and

3. by cutting the timber in such a way as to prevent the perpetuation of the forest on the sustained yield basis contemplated by the 1908 Act.

(Chapman 1994, 38)

At first this judgment seemed like an enormous victory for the tribe. Tribal members had complained about the BIA's management of the forest for years. Tribal membership and leadership efforts had finally paid off. Good times were ahead. With the court settlement the Menominee became one of the nation's wealthiest tribes. Unfortunately, however, the settlement did not lead to more prosperity, but led to the federal Termination policy disaster.[1]

The forestry plan implemented in 1951 concurrent with the Menominee court victory was simple. As Newman said, it was "little more than a cutting budget" (Newman 1957, 78), but it was also flexible, designed to change with changing economic conditions and forest problems or opportunities that might present themselves during the plan's implementation.

The plan's major problem was its strict adherence to the rule calling for an annual harvest of twenty million board feet. The BIA, stung by the Court of Claims decision, was determined not to repeat past mistakes by ignoring the LaFollette Act again. This caused another serious management error as the concluding opinion for the case, *Menominee Tribe of Indians et al. v. The United States of America*, filed during Termination, stated:

> From the time of the publication of the Stand and Structure Analysis dated January 1, 1952, the defendant [The United States of America] knew or should have known that the 20 MBF harvest limitation was the principal and controlling cause of substantial underproductivity in the Menominee Forest.
>
> As trustee of the forest, the defendant was obligated to be cognizant of existing forest conditions, and to inform Congress of the need to increase the harvest ceiling.
>
> The injuries to the Menominee Forest were perpetuated into the post-termination period by the 1961 Management Plan. (*Menominee Tribe of Indians et al. v. The United States of America* 1984, 329)

In 1954 Congress passed the Termination Act. In the strictest sense, Termination, although it caused great suffering, did not harm the forest's overall health. Most of its effects were economic. But in a more global sense, the act struck at the heart of the Menominee idea of sustainability. The Legend Lake development near Keshena removed acreage from the forest and passed those acres into non-Menominee hands, violating in the deepest sense possible the Menominee land ethic. Termination also made operating Menominee Tribal Enterprises more difficult by draining cash from sawmill operations for the payment of Wisconsin taxes.

By the time of Termination, all parties affected by the act realized the forest's long-term value. The act acknowledged this value by requiring that the tribe's Termination plan had to contain provision for the forest's protection on a sustained yield basis. The plans also required protection of the water, soil, and wildlife. It stated that "The sustained yield management requirement contained in this Act . . . shall not be construed by any court to impose a financial liability on the United States (United States Congress, PL 83-399, 1954). In other words, you are mature enough to manage your own affairs, Menominee. We are granting you the same property rights we would grant any other U.S. citizen as long as you manage your forest in a manner that protects the interest of the United States public. This may cost you, but even if it does and even if this provision turns out to be unfair, this law denies you any legal recourse.

The deed that finally transferred forest control to Menominee Enterprises, Inc. contained additional restrictive language that recognized Wisconsin's interest in preserving what the State saw as a major public asset:

THE PARTIES HERETO MUTUALLY COVENANT and agree for the benefit of the State of Wisconsin as follows:

1. That the lands conveyed hereby shall be operated on a sustained yield basis until released therefrom under the laws of Wisconsin or by an act of Congress.

2. That for a period of 30 years commencing with the date of this deed the ownership of lands conveyed hereby shall not be transferred, nor shall such lands be encumbered without prior consent of the State Conservation Commission of Wisconsin unless released from sustained-yield basis under the laws of Wisconsin. (*Menominee Tribe of Indians et al. v. the United States of America* 1984, 101a–102a)

These requirements reversed the more flexible, modulated approach to forest management slowly evolving up until Termination. They also, at a crucial moment in Menominee history, harmed the value the Menominee could realize from timber harvesting and manufacturing operations as a result of the arbitrary, and by this point in time unjustified, twenty million board feet per year cutting limitations. Wisconsin foresters became integral to the forest's management since special tax considerations were essential to maintaining the public interest in the forest. Wisconsin, the federal government, and the tribe all recognized that normal taxation would bankrupt the operations of Menominee Enterprises, Inc. in a short time period. Wisconsin's foresters, along with MEI's foresters, were rigorous in their applications of sustained yield principles without regard to economic considerations. The result of these applications, in Newman's words, was that "more refined techniques in all phases of forestry made possible the development of a more intensive and sophisticated program on the forest." The Enterprise could not sell maple when maple had a high market value if the management plan restricted the maple harvest for that year. The Enterprise therefore lost money. At the same time forest management techniques and applications became more sophisticated.

None of this is meant to convey that the Menominee disapproved of the restrictiveness imposed by sustained yield harvesting techniques. The opposite was true. During negotiations leading to the final date when termination was put into effect, the Menominee had approved of, and in some cases lobbied for, strong sustained yield restrictions. Menominee leaders such as George Kenote had every intention of preserving the forest with as much skill as the Enterprise could call upon. Kenote championed Legend Lake land sales and battled DRUMS and Menominee Restoration, causing future generations of Menominee to remember him with more than a touch of asperity, but at the onset of Termination his approach to forest management agreed with the restrictions imposed by both Congress and the State of Wisconsin.

Among the innovations of the fifth period were:

- The replacement of A-frame loaders by portable loaders early in the period with the subsequent replacement of the portable loaders by loading units on trucks that could easily be moved through the woods.

- The introduction of power saws to replace the large two-man saws and axes.

- The expansion of the cut from twenty million board feet per year to the amount of annual growth (net growth) derived from increasingly sophisticated growth data.

- The introduction of aerial photography to establish volume, area, and expected yield information on the forest.

- The continuing evolvement of CFI (Continuous Forest Inventory) methodology, establishing 950 one-fifth-acre plots post-stratified according to timber type in order to increase both accuracy and systemization of the process.

The most important innovation was in how CFI data was utilized. As forest plots were tallied, trees were categorized into four cutting classes that were then prioritized. Trees in classes one and two were considered, on the basis of risk and vigor, to have a high cutting priority. CFI data established class one and two trees throughout the forest, thus resulting in a high cutting priority volume. This volume was then increased by the total volume of overstocked trees, and this sum was divided by fifteen, the number of years in the cutting cycle. This number then set the annual allowable cut of 30.297 million board feet.

Other important developments during the period were a dramatic planting program expansion, leading to the planting of the one millionth tree in 1959, improvements in the treatment and prevention of disease and insect infestations[2], and improved communications that helped with fire prevention and the discovery of disease. Tribal members also became increasingly cooperative in helping to preserve the forest as the fifth period continued. The reservation's sustainable development ethic strengthened. Increasingly, Menominees wanted economic development, but they also wanted a pristine environment surrounding them. They were not interested in destroying their environment in pursuit of jobs even though reservation poverty increased throughout the period.

The Sixth Period:
From Restoration to the Present

The sixth period of Menominee Forest use began when Richard Nixon signed the Menominee Restoration Act (United States Congress, PL 93-197) on 22 December 1973. In many ways the start of the Menominee Restoration Committee's work marks the first point in Menominee history occurring after the Americans assumed control over the Great Lakes territories that the Menominee had dominion over their own land and lives. They lived in the midst of a powerful non-Indian society, but now they had the opportunity to plot a future independent of the non-Indian world. During Termination the First Wisconsin Bank had controlled MEI, and thus most Menominee land. Before Termination either the Indian agent's office, the Bureau of Indian Affairs, or the U.S. Congress dictated the disposition of Menominee Forest resources. The Menominee had fought what they believed was wrong in their world year after year, but not until the Restoration Committee was seated could they truly say their decisions about the forest and their lives were preeminent.

At the heart of current Menominee forestry practices is CFI data that allows Menominee foresters to measure changes in the forest.

These measures help determine the long-term impact of forest management policies. Management decisions ranging from silvicultural treatments to log grade/lumber recovery for Menominee Tribal Enterprises (the former MEI) result from analysis of this data.

The current management plan contains silvicultural prescriptions for fourteen cover types and describes in detail the cutting and/or management options available to improve tree-stand quality and volume. Strict silvicultural prescriptions specify the minimum stocking levels of each species before any green, standing timber is cut. Understocked acreage is allowed to grow and develop until stocking levels exceed silviculturally prescribed minimums. Cut-leave determinations for each timber stand are based on the CFI for "each fixed radius, systematically located plot."

> The species/cover-type harvest prescription is applied on each plot to determine the forestwide removal, and is based on the excess stocking of fully stocked stands, not on the net growth of all stands. This method ensures that the silvicultural prescription controls the volume removed, and that net growth on under-stocked stands does not contribute to the annual allowable cut. (Pecore 1992, 15)

The CFI is currently divided into 109 distinct compartments. Compartment management is still based on a fifteen-year cutting cycle for timber types subject to all-aged management. All remaining cover types are cut as closely as possible to the compartment schedule established by the management plan. The result is a reasonably even flow of sawtimber and pulpwood to the Neopit mill. The management plan also allows "weather conditions, insects and disease, market disruptions, and the attainment of annual allowable cut in a compartment" to change harvesting schedules.

The Menominee have also founded the Menominee Sustainable Development Institute at the College of the Menominee Nation. The institute is working toward both a model of Menominee sustainable development and an increased understanding of the forest's ecology. As part of this effort a small mammal study and other biodiversity studies are being planned, and an effort is being made to provide tours to forest professionals who want to learn about Menominee management of their forest and other resources.

As Larry Waukau, MTE's current president, was reported as saying recently, the Menominee have lived through three hundred years of contact with the Europeans and survived. They have also managed their forest in such a way as to be able to harvest more timber today than was available for harvest during earlier periods. Now, according to Waukau, the task is to move forward into an era where secondary processes increase the forest's economic benefit to the Menominee. On the planning boards is a small mill designed to reduce the amount of waste by-products from harvesting and improve handling of smaller diameter logs. A number of other secondary process and finished product manufacturing are being studied. Mill management is also working hard at improving workforce and mill efficiencies by applying continuous improvement management techniques to mill operations (Waukau 1995, conversation).

The Importance of the Human Use of Land

In his book *The Unsettling of America Culture and Agriculture*, Wendell Berry makes the point that how we use land is extremely important. He defends the preservation of wilderness on three grounds: 1) for its value as a memory of both our biological and cultural roots in nature, 2) for its value in teaching us the arts of humility and of leaving some places alone, and 3) for its value as a standard of civilization and a cultural model. "Only if we *know* how land *was* can we tell how it *is*," he states. But he points out that "we cannot hope for purposes practical and humane, we cannot even wish—to preserve more than a small portion of the land in wilderness." In the end, according to Berry, we need to know how to use the land before we can learn how to govern ourselves and our culture (Berry 1977, 28–31). If we do not know how to use the land then we cannot hope, over the long-term, to preserve either ourselves or any culture.

The Menominee Forest has been intensively logged since the 1908 passage of the LaFollette Act. Still, its sawtimber volume has increased rather than decreased. This is the opposite of what has happened in other forests of the Great Lakes region. A great diversity of both species and ages of trees within species exist in the forest. Soil and the abundant water resources of the reservation are ecologically healthy. Menominee land use has been both kindly and good and is worthy of study and emulation.

The history of the Menominee Forest is complicated by the Menominee relationship of the tribe with the federal government, but it helps to tell us how to use the land. It recounts mistakes as well as triumphs. It helps to tell all of us how to develop and preserve the best in our culture.

8

Managing the Forest and Menominee Ecosystems

Sustaining the Forest

THE RESERVATION'S POST-RESTORATION political, institutional, cultural, social, and economic systems have created a powerful set of sustainable development policies, guidelines, and procedures. These are contained in the *Menominee Tribal Enterprise Forest Management Plan* that will be the primary management document used from 1996 through the year 2005 (Menominee Forestry Department 1995). One of the plan's most significant aspects was that the Bureau of Indian Affairs and Wisconsin Department of Natural Resources, in an intense multi-agency effort, helped with all aspects of the planning. This spirit, where the Enterprises and its forestry department supplements its expertise with federal and state government expertise, is an indicator of how sophisticated the Menominee have grown in pursuit of their forest-preservation mission.

The concept of "sustainable forestry" outlined in the plan shows how far the Menominee have thought through the process of achieving sustainable development. In past plans the focus was invariably on the forest, its preservation, and its short-term and long-term productivity. The 1996 Plan calls for a "forestry that is ecologically viable, economically feasible, and socially desirable." This forestry is different from that practiced in the past because it brings under its umbrella "not only . . . forestry products and social benefits, but also . . . wildlife, site productivity, and other ecosystem functions." It contains the "three components of a sustainable system" (Menominee Forestry Department 1995, 4):

First, it must be sustainable for future generations. . . . Second, the
forest must be cared for properly to provide for the needs of the people. Management must conserve the productive capacity of the land
to produce forest products in order to sustain the tribe's economy.
Third, (it must) maintain the forest's diversity." (Menominee Forestry
Department 1995, 4)

Additional goals listed include using the forest environment as a
benchmark to "evaluate the effectiveness of ecosystem management
strategies," maintaining individual, structurally complex stands of
trees "containing seedlings, saplings, and sawtimber trees, as well as
dead material in various stages of decomposition," and allowing a
diversity of "wildlife, invertebrates, and fisheries." This will allow the
maintenance of adaptability amidst the myriad threats, insects, fires,
disease, and pollutants that can negatively impact the forest. It will
also minimize "societal and ecological effects of an unpredictable
future" (Menominee Forestry Department 1995, 5).

The plan additionally calls for maintaining the Continuous
Forestry Inventory (CFI). CFI data will be used to set the annual
allowable cut's size so that the cut "will not diminish the volume or
value of the forest as a whole." It will establish silvicultural prescriptions designed to maintain the forest's diversity and health as well as
to restore forest sections damaged during the pine baron days and the
"Ayer Report" era when clear-cutting and other undesirable forestry
practices were common.

Issues of public awareness, education of Menominee and other
school children, the balance between sustainability and market economics, and challenges to sustainability are discussed in the report
with a depth this chapter cannot cover. Instead, key data, concepts,
and issues the Menominee believe are important enough to address in
the plan are discussed. What can be gained from the data is a current
Menominee Forest snapshot. Concepts and issues addressed show
how a society and culture centered in the idea of sustainability thinks
about sustainability. Sustainability may be forever, but forever has to
be worked out within everyday life's timetables.

Current Data About the Menominee Forest

One worry facing Menominee foresters is that since 1963 the total forest
acres available for "sustained yield management" has been declining.
The table 8-1, "Total Acres by Forest Cover Type," illustrates this trend:

Table 8-1
Total Acres by Forest Cover Type

Cover Type	Total Acres by Forest Cover Type			
	1963	1970	1979	1988
Red Pine	2,728	1,736	3,226	5,460
White Pine	15,388	17,622	19,359	16,629
Jack Pine	3,474	3,474	2,234	744
Swamp ConiFers	22,337	22,584	22,089	21,344
Hemlock-Hardwoods				
Hemlock	5,459	4,467	4,964	3,971
Hemlock-Sugar Maple	3,971	4,219	4,219	4,963
Hemlock-Yellow Birch	15,388	15,883	12,905	11,644
Northern Hardwood				
Sugar Maple-Beech	28,789	28,541	35,987	43,679
Mid-tolerant Hardwoods	53,360	61,551	61,303	61,800
Red Oak	3,475	4,220	4,714	5,957
Swamp Hardwoods	2,731	2,977	2,481	1,985
Aspen/White Birch	47,901	39,957	34,003	25,315
Pin Oak	9,925	8,191	8,437	7,445
Temp. Nonproductive	3,723	3,475	3,225	7,197
Perm. Nonproductive	14,891	14,643	14,394	15,387
TOTAL	233,540	233,540	233,540	233,540

(Menominee Forest Department 1995)

The trend is set by the two bottom rows, temporary nonproductive and permanent nonproductive acreage. Temporary nonproductive acreage is less of a concern than permanent nonproductive acreage, but an increase in both categories has occurred and is occurring. From 1979 to 1988 a total of 993 acres were permanently lost to the forest, primarily because of other reservation development activities. This may not seem like much, but in a society like Menominee any loss is both worthy of comment and a cause for concern. The foresters have held several meetings with reservation economic development and legislative groups in the hope of stopping the continual nibbling at the forest. These meetings have been well received. The development of Middle Village in fields purchased from non-Indians beyond the reservation's borders was a response to concerns about maintaining the forest, but population and development pressures continue. Tribal vigilance will be necessary if its sustainable development record is to continue.

This decrease in acreage has not resulted in either a loss of saw-timber or cordwood volume. The forest has shown a dramatic increase in growth between 1979 and 1988. This is established in tables 8-2, 8-3, and 8-4.

Table 8-2
Sawtimber Board Foot Volume by Species, 1963–1988

Species	Board Foot Volumes			
	1963	1970	1979	1988
Balsam Fir	852,731	853,702	1,775,083	1,985,201
Tamarack	743,407	1,379,057	2,191,460	2,639,663
White Spruce	3,826,358	2,998,903	4,098,030	5,104,804
Black Spruce	765,272	1,138,270	1,227,218	676,277
Jack Pine	3,345,330	3,524,258	5,018,443	3,425,018
Red Pine	25,603,798	29,485,561	34,888,043	38,373,289
White Pine	314,832,695	309,696,890	327,864,331	347,933,815
Cedar	36,055,220	40,364,791	48,387,437	52,182,437
Hemlock	360,377,282	352,206,882	349,844,674	348,762,800
Red Maple	17,010,892	22,371,376	33,069,131	45,485,109
Sugar Maple	257,393,603	257,008,142	300,405,337	354,478,435
Yellow Birch	107,793,957	103,188,517	95,372,339	86,607,138
White Birch	5,007,062	6,982,846	11,877,713	13,569,179
Hickory	2,077,166	1,816,853	2,761,240	3,708,619
Beech	28,249,451	27,165,242	35,326,335	40,380,305
White Ash	5,444,360	5,384,891	8,152,231	11,060,408
Black Ash	8,942,744	9,259,385	9,576,680	9,707,853
Butternut	437,298	700,474	986,157	894,431
Balsam Poplar	1,727,327	1,860,633	2,826,983	3,185,048
Bigtooth Aspen	1,705,462	3,655,597	9,204,132	15,816,165
Quaking Aspen	40,384,470	45,355,666	51,455,481	53,033,237
Black Cherry	568,487	1,422,837	2,564,008	3,206,864
White Oak	634,082	1,160,159	2,564,008	3,643,172
Pin Oak	17,623,109	20,269,955	27,437,079	34,970,086
Red Oak	66,666,080	78,628,162	100,807,160	126,114,827
Basswood	95,637,073	95,658,426	106,088,579	105,150,228
Soft Elm	76,592,745	70,857,283	10,343,691	872,616
Hard Elm	42,024,338	39,598,648	8,941,157	305,416
Miscellaneous	0	0	0	87,262
ALL SPECIES	1,522,321,799	1,533,993,406	1,595,054,160	1,713,359,702

(Menominee Forest Department 1995)

Tables such as these are used in several different ways by the foresters. Table 8-2, for instance, shows a problem with elm board foot volume on the reservation. The soft elm volume went from 76,592,745 sawtimber board feet in 1963 to just 872,616 board feet in 1988. A similar loss was experienced by hard elm. From 42,024,338 board feet in 1963 to only 305,416 board feet in 1988. In this case Dutch Elm disease resulted in the loss of thousands of board feet per year.

The major point this table makes, however, is that the sawtimber total board feet volume of the forest rose from 1,522,321,799 in 1963 to 1,713,359,701 in 1988. This increased volume was achieved in spite of smaller forest acreage.

Table 8-3
Cordwood Volume by Species, 1963–1988

Species	Volume in Cords			
	1963	1970	1979	1988
Balsam Fir	41,366	44,071	45,119	39,189
Tamarack	19,134	25,173	31,698	26,211
White Spruce	12,990	10,275	11,850	13,945
Black Spruce	22,594	23,682	21,591	18,064
Jack Pine	48,500	54,609	62,253	26,570
Red Pine	36,202	45,761	61,093	79,562
White Pine	366,734	373,263	409,983	426,786
Cedar	350,349	386,683	445,948	476,782
Hemlock	791,104	780,986	299,578	384,048
Soft Maple	155,258	200,121	770,247	765,127
Hard Maple	670,975	711,691	873,218	1,061,476
Yellow Birch	296,146	310,715	332,279	340,228
White Birch	83,267	104,715	139,680	151,125
Hickory	19,139	19,696	23,829	26,961
Beech	79,510	81,043	101,852	114,973
White Ash	30,833	35,966	49,128	60,379
Black Ash	60,445	63,771	68,291	71,139
Butternut	3,855	3,750	3,897	2,840
Balsam Poplar	10,036	9,121	10,666	11,745
Bigtooth Aspen	32,101	48,014	71,902	96,674
Quaking Aspen	448,074	460,970	427,732	369,210
Black Cherry	12,354	17,341	24,536	23,293
White Oaks	8,466	10,653	14,494	18,663
Pin Oak	110,608	122,869	156,528	179,222
Red Oak	201,027	231,136	280,914	317,635
Basswood	240,368	256,369	300,641	323,319
Soft Elm	160,167	159,976	41,268	9,490
Hard Elm	69,227	65,268	21,665	2,163
Miscellaneous	8,527	8,368	8,766	9,739
ALL SPECIES	4,389,356	4,666,056	5,110,646	5,446,558

(Menominee Forest Department 1995)

Table 8-3, "Cordwood Volume by Species 1963-1988," illustrates this point. As shown in table 8-2, "Sawtimber Board Foot Volume by Species 1963-1988," a steady increase in cordwood volumes occurred between 1963 and 1988—in this case a volume growth for all species from 4,389,356 in 1963 to 5,446,558 in 1988. Again, this chart helps to define what is happening in the forest as volumes for individual species rise and fall.

Table 8-4, "Sawtimber Volume Growth by Species, 1971-1988" establishes in a more precise way the forest's growth volumes. This table shows the increases or decreases in board foot growth per acre, per year, per species and also shows the total board foot growth per species, per year. These growth rates are averages, but they again show that the total growth rate of board feet in the forest is increasing, although, as noted, from 1971 through 1988 the elm production in the forest decreased because of Dutch Elm disease. Black spruce also developed problems.

Table 8-4

Sawtimber Volume Growth by Species, 1963-1988

Species	Board Foot Per Acre Per Year			Total Board Foot Growth Per Year		
	1963–70	1971–79	1980–88	1963–70	1971–79	1980–88
Balsam Fir	0	0.5	0.1	0	109,571	21,815
Tamarack	0.4	0.4	0.2	87,559	87,658	43,631
White Spruce	0.2	0.6	0.5	43,780	131,488	109,077
Black Spruce	0.3	0	-0.3	65,669	0	(65,446)
Jack Pine	0.5	0.8	0.2	109,449	175,317	43,631
Red Pine	2.5	2.9	2.9	547,245	635,523	632,647
White Pine	23.6	26.2	27.8	5,165,993	5,741,625	6,064,681
Cedar	2.8	4.1	2	612,914	898,499	436,308
Hemlock	27.2	24.2	25.3	5,954,026	5,303,333	5,519,296
Red Maple	4.5	6.3	8.2	985,041	1,380,620	1,788,863
Sugar Maple	26.8	35	41.1	5,866,466	7,670,110	8,966,129
Yellow Birch	4.6	4.1	0.9	1,006,931	898,499	196,339
Paper Birch	2.2	2.6	1.4	481,576	569,780	305,416
Hickory	0.5	0.6	0.7	109,449	131,488	152,708
Beech	3.1	5.2	4.7	678,584	1,139,559	1,025,324
White Ash	0.9	2.2	2.2	197,008	482,121	479,939
Black Ash	0.2	0.2	0.1	43,780	43,829	21,815
Butternut	0.2	0.1	0.1	43,780	21,915	21,815
Balsam Poplar	0.3	0.5	0.2	65,669	109,573	43,631
Bigtooth Aspen	1.3	2.8	3.7	284,567	613,609	807,170
Quaking Aspen	11.3	10.7	8.8	2,473,547	2,344,862	1,919,755
Black Cherry	0.6	0.6	0.6	131,339	131,488	130,892
White Oaks	0.3	0.7	0.5	65,669	153,402	109,077
Pin Oaks	3.4	4.1	6.3	744,253	898,499	1,374,370
Red Oak	10.5	13.2	18.6	2,298,429	2,892,727	4,057,664
Basswod	10.3	12.5	9.2	2,254,649	2,739,325	2,007,017
Soft Elm	5.6	-9	-2.7	1,225,829	(1,972,314)	(589,016)
Hard Elm	2.8	-2.7	-2.3	612,914	(591,694)	(501,754)
Miscellaneous	0	0	0	0	0	0
TOTAL VOLUME	146.9	149.4	161	32,156,115	32,740,412	35,122,794

(Menominee Forest Department 1995)

These tables clearly show how successful the Menominee foresters have been in increasing the forest's productivity. Individual species have shown signs of one kind of stress or another, usually coming into the forest from beyond the reservation's borders, but even in the case of elm and black spruce problems, enough mature trees are left in the forest to regenerate the species over time.

Overall Forest Management Policy

The reasons for the foresters' management success is straightforward. Menominee foresters use different management techniques today than were used by earlier foresters. This difference is attributable to an increasing sophistication driven by scientific and technological advances, but some of the difference also results from discussions between the foresters and those engaged in mill, logging, or other forestry operations.

The most troubling aspects of the forest's history to most Menominee is that portions of the forest were either damaged or destroyed at different times. There has long been a desire to restore the historical character of the forest. The idea is to ensure the mixes of tree stands become similar to those that existed before logging started. Behind this thought is the idea that the forest developed the way it did over thousands of years until, in its climax state, it had reached a sustainable state, resistant to disease, natural catastrophe, and a variety of interferences—even those caused by humans. The Menominee foresters believe that by restoring the forest's original character, they can help ensure both greater productivity and long-term sustainability.

More than 9,000 distinct tree stands exist on the 220,000 acres of reservation forestland. The management plan defines the significance of tree stands as follows:

> These stands are defined by attributes such as tree species composition, tree size, volume, or number of trees per acre. The forest stands occur on a variety of soils and topographical or geologic features interspersed with streams and lakes. This combination of physical and biological elements has evolved into diverse plant and animal communities (or ecosystems). The management and protection of Menominee Forest ecosystems require forest management practices which maintain the environmental conditions necessary to ensure diverse plant and animal populations (biodiversity). (Menominee Forestry Department 1995, 21)

Traditionally, during earlier periods, foresters made management decisions about the forest based upon the condition and appearance of various tree stands. Timber stands "were harvested and regenerated with the objective of maintaining their existing species composition." Foresters usually had little or no knowledge "of past historical events such as cutting practices, fire, windthrow, or land use (grazing)." Still, these historical events affected forest productivity. This meant that prescriptions used to treat the forest often categorized what could be highly productive forest land as a less productive site simply because past events had caused inappropriate species to grow where they would not have grown if the forest had never experienced these events.

An example would be "aspen growing on a site suitable for high value sugar maple." One cutting prescription, in this example, would regenerate Sugar Maple and other high quality hardwoods on the site. Another cutting prescription could regenerate aspen. The forester using traditional management techniques would look at the site and try to regenerate the present species. Today's Menominee forester would attempt to regenerate the favored species (Menominee Forestry Department 1995, 21).

Current management decisions are based upon the Forest Habitat Classification System which "provides a method to accurately assess forest site productivity." It also "provides a guide that can indicate which forest cover type is best suited" to an individual site where a tree stand is located. This is achieved without regard to either the species or tree quality currently growing on a site. The concept behind the classification system is that understory plants, ranging from grass to bushes to wildflowers, are better indicators of the kinds of trees that should be growing than are the trees present. Certain understory types can usually be found living in the presence of certain tree species. Tree species have been dislocated due to historical harvesting practices. Understory types have, for the most part, never been dislocated or have migrated back to a favorable environment. Therefore, by identifying understory types present, the forester can establish what species of trees are most likely to flourish on that site. Studies completed on the Menominee Forest have identified "eleven understory plant associations . . . across the spectrum of sites, ranging from the dry, nutrient poor sites (pin oak, jack pine) to the moist, nutrient rich sites (sugar maple, basswood)." The management plan states:

> While understory plant associations can be identified with each habitat type, tree species are less specific in their ecological range. For this reason, single tree species will grow on a number of different sites (habitat types), but may achieve its best form and quality on only one

or two habitat types. Therefore, just because a tree species is found growing in a stand does not mean that it will be productive (reach large size and quantity) on that site. (Menominee Forestry Department 1995, 21)

None of this means that a species could not naturally become reestablished on a site that it had been harvested from at some point in the site's history. Over a period of time this has happened when a seed source is present and a favorable combination of chance events allows it to happen. What the Menominee foresters are attempting to do is to use the Forest Habitat Classification System to help "identify which tree species are best suited to a given habitat type" and then to develop "a silvicultural prescription which either maintains or establishes the most productive tree species" on each forest site. Three criteria are used to help identify which species would be the most "productive" on any one site:

1. The sawtimber potential from both quality and quantity measures;

2. The species' biological/ecological suitability to the site;

3. The competitiveness of a species with the other tree species currently associated with it.

(Menominee Forestry Department 1995, 22)

These criteria mirror the values stated by Menominee foresters again and again. First, pay attention to the marketplace and the needs of the Menominee people, but, second, at the same time, and more importantly, pay attention to the marketplace only within the context of biological/ecological systems. Third, make sure that silivicultural prescriptions pay attention to how the forest is currently constituted and what is healthy for the forest's needs. Make changes, but do so cautiously, always aiming toward long-term forest health and productivity, not short-term health or productivity.

When the foresters decide to change species mix on an individual site, the tree species that become the management target are called an objective species. Associate species "are those species likely to exist as a result of the efforts of silvicultural systems designed to promote the objective species." Individual tree species that tend to associate together are a community. A community of trees "with a specific combination of or abundance of species" is called a cover type. Cover types are managed in units called stands (Menominee Forestry Department 1995,).

In order to get to the point where a silviculture prescription for an individual site is decided upon, a full silvicultural review of the forest stand is made. During the review three management alternatives are considered:

1. If the stand primary type is currently a featured forest cover type, the objective will be to maintain the present cover type through harvesting practices.

2. If the current stand includes a minor component of one or more featured species that could become the major component of the stand through management, the management prescription for this stand will be tailored to increase the presence of this featured species.

3. If no featured forest type or species is present, establish a featured cover type through a seeding or planting effort.

(Menominee Forestry Department 1995, 22)

These efforts are undertaken to make the forest a more complex structure and increase the forest's overall productivity.

Decisions based on these criteria are always made with the understanding that the forest is, in Dan Pubanz's words, "a highly complex entity whose processes remain, for the most part, unknown." Silvicultural prescriptions can be developed and applied, but the "same management practice can give different results on different sites which confounds long-term management" (Pubanz 1995, 239). In order to better manage this complexity, the foresters "have classified all of Menominee's 884 Continuous Forest Inventory (CFI) plots."

As a result, we are able to analyze thirty years of inventory data by habitat type and are now able to perform many specific analyses for each habitat type. Tree species growth, "competitiveness" relative to other species, grade, age at maturity, cull rate, and regeneration status can all be assessed by habitat type. This has provided a wealth of insight into the biological functioning of forest trees. . . .

These analyses of CFI data by habitat type, in combination with the work of Dr. John Kotar on tree and habitat type relations, allowed us to determine which tree species were best suited to which habitat types given our management goals. (Pubanz 1995, 241)

Such intensive research has not always guaranteed the results the foresters are striving for, but it has reduced some of the concern caused by the forest ecosystem's underlying complexity.

A description of the eleven habitat types found on the reservation follows on the next three pages of tables.

Silvicultural Prescriptions

Silivicultural prescriptions are applied to the forest based upon the way various stands of trees best reproduce. Silvicultural prescriptions available for managing the stands found in each habitat type include the following:

CLEARCUTTING: In the clearcutting method the area is cut clear in the literal sense of the word; virtually all the trees, large or small, in the stand are removed. This is the meaning of "clearcutting" in its narrowest sense and in its usage as a technical silvicultural term. However, it is also loosely applied to any type of cutting in which all the merchantable timber is cut and all trees that cannot be utilized profitably are left. Reproduction of the stand is secured after the cutting either naturally or artificially by seeding or planting.

Clearcutting works best for tree species adapted to fire or other disturbances. These species are shade intolerant and need full sunlight to regenerate. Red pine, jack pine, and aspen respond well to clearcutting.

SEED TREE: The seed tree method removes all but a few selected trees. These selected trees are called seed trees. The uncut trees ensure a seed source for regeneration. Good seed producers with the best genetic attributes are selected to become the seed trees. After a new stand of trees is established these seed trees may be either removed in a second cutting or left standing.

Like the clearcutting method, the seed tree method works best for tree species that are shade intolerant. White birch is a species that responds well to the seed tree method. This is not a common harvesting method on the Menominee Forest, although it has been used on rare occasions.

SHELTERWOOD: The shelterwood harvesting method removes portions of a forest stand gradually, over time. As the name of the method

Table 8-5
Menominee Habitat Type Descriptions

Habitat Type	Common Names	Reservation Location	Soil Characteristics	Floral Characteristics
QV	Pin Oak, Blueberry	Identified only in southeastern corner	Soil: deep, droughty outwash sands Nutrients: poor Moisture: very low	Presence of pointed-leaved tick trefoil, early meadow rue, princess pine, and pyrolas
PMV(Q) Pinus/Maianthemum Vaccinium (Quercus)	White Pine, Wild Lily-of-the-Valley, Blueberry	Found almost exclusively in the eastern half	Soil: most loamy sands Nutrients: medium Moisture: low (not as dry as QV)	Presence of witch hazel, American fly-honeysuckle, downy yellow violet, beech seedlings, and white ash seedlings
AQVib Acer/Quercus/Vibernum	Sugar Maple, Red Oak, Maple Leaf Vibernum	Found in eastern half	Soil: sandy loams and loams Nutrients: medium Moisture: medium	
AQVib(Ha) Acer/Quercus/Vibernum (Hamamelis)	Sugar Maple, Red Oak Maple Leaf Vibernum (Witch Hazel phase)	Found in the eastern half, with close proximity to ACVib	Soil: mostly loams and fine sandy loams of outwash, overlying reddish-brown till. Nutrients: medium Moisture: medium (more moisture than AQVib)	Presence of wild geranium, sweet cicely, bristly green brier, spinulosa, shield fern, and alternate-leaved dogwood. Has well-developed shrub layer

Habitat type	Dominant species	Distribution	Soils	Indicator plants
AFVib Acer/Fagus/Viburnum	Sugar Maple, American Beech, Maple Leaf Viburnum	Found in eastern half, with close proximity to AQVib and AQVib(Ha)	Soils: most common soils are loam and fine sandy loams over red till. Nutrients: rich. Moisture: medium (more moist than AQVib(Ha))	Presence of maidenhair fern, Indian cucumber root, and spikenard. Doesn't have a well-developed shrub layer. Hazelnut, bush honeysuckle, and alternate-leaved dogwood have a combined cover of less than 15 percent
TMC Tsuga/Maianthemum/Coptis trifolia	Hemlock, Wild Lily-of-the-Valley, Goldthread	Found in low-lying areas, within many other habitat types	Soils: no specific soil type dominates. Nutrients: medium. Moisture: high	Presence of wood sorrel, bunchberry, and goldthread
ATM Acer/Tsuga/Maianthemum	Sugar Maple, Hemlock, Wild Lily-of-the-Valley	Found scattered throughout	Soil: fine sands. Nutrients: medium. Moisture: medium	Presence of round-lobed hepatica, chokecherry, trilliums, wood anemone, bristly green brier. Absence of wood sorrel, bunchberry, and goldthread
ATFD Acer/Tsuga/Fagus/Dryopteris	Sugar Maple, Hemlock, Beech, Shield Fern	Found in northeastern part and scattered between Neopit and Keshena	Soil: fine sandy loams. Nutrients: medium to rich. Moisture: medium	Presence of Indian cucumber red elderberry, sweet cicely, gooseberries, and jack-in-the-pulpit

Table 8-5 (continued)

Habitat Type	Common Names	Reservation Location	Soil Characteristics	Floral Characteristics
ATDH Acer/Tsuga/Dryopteris/ Hydrophyllum	Sugar Maple, Hemlock, Shield Fern, Virginia Waterleaf	Western half	Soils: fine sandy loams Nutrients: medium to rich Moisture: rich	Presence of maidenhair fern, bloodroot, sharp-lobed hepatica, leatherwood, blue cohosh, and Virginia waterleaf
AFDAd Acer/Fagus/Adiantum	Sugar Maple, Beech, Maidenhair Fern	Central portion	Soil: loams Nutrients: rich to very rich Moisture: rich	Presence of beech seedlings, bitternut hickory seedlings. Absence of clubmosses, starflower, yellow beadlily, oak fern, round-lobed hepatica, and mountain maple
AH Acer/Hydrophyllum	Sugar Maple, Virginia Waterleaf	Western half, intermixed with ATDH	Soil: silt loams and loams Nutrients: rich to very rich Moisture: rich to very rich	Presence of early meadow rue, wild leek, Canadian white violet, wood nettle, and the absence of beech. Coverage of Virginia water-leaf, wood nettle, and sweet cicely is 15 percent or greater

(Menominee Forestry Department 1996)

implies, selected trees that are not cut serve as a shelter and a seed source for the next generation. Tree species that grow well in a shelterwood system can regenerate in partial shade.

There are two steps in the shelterwood method. The first step is the seed cut. This step removes about 50 to 60 percent of the mature trees. The intent in this cut is to remove the weaker trees and open the forest canopy enough to allow sunlight through so new seedlings can sprout. The second cut is the overstory removal cut. This cut, made after the establishment of new seedlings, removes most of the remaining mature trees. A few trees remain as insurance against a natural disaster and for aesthetic reasons. White Pine are an example of a species that responds well to the shelterwood method.

SELECTION: The selection method is what many people consider "ideal" forest management. Using this method foresters consider each individual tree for removal. Trees identified for harvesting are past maturity, have a poor quality, or are not likely to survive until the next scheduled cutting. Selection cutting works best for trees that can regenerate in full shade. Sugar maple, beech, hemlock, and basswood reestablish easily when managed by selection. The selection method results in a forest with all ages of trees.

Selection cutting is the most difficult silvicultural treatment to apply. Enough trees need to be removed to provide room for new trees. If too many trees are cut, the remaining trees become poorer in quality. They tend to grow too many limbs and become what is known as 'limby.' This reduces their sawtimber value.

(Menominee Sustainable Development Institute 1996, 20–21)

Some of these methods have been controversial with some environmentalists in the past. Clearcutting prescriptions cause especially strong reactions. The word clearcutting causes visions of immense numbers of acres devastated by giant machines that have chewed through the forest like a giant omnivorous insect. Left standing in this destruction are brush, the upturned circles of roots of toppled trees, and an occasional straggly-looking sapling that has miraculously escaped the disaster.

The point Menominee foresters make when confronted with such reactions is that certain species respond best to this prescription. These species tend to regenerate quickly, and, if you want to maintain diversity within the forest while still harvesting, then the goal should be tree growth and not the satisfaction of environmental purists. Ecology tells us that complex systems are often healthy systems. The idea that one type of management can fit all the complex ways plants and animals have developed in order to survive is not realistic. Multiple management methods need to be applied based upon the best scientific information and site data available. This information and data needs to be continually reviewed, and its application has to be conservative, based upon past experience and an intimate knowledge of the forest.

Supplementing the silvicultural harvesting prescriptions are other tools important to the forest's reproduction processes. One of the most important of these is prescribed fires. Many of the forest's cover types regenerate following large-scale disturbance. As was described earlier, windstorms, insect and disease outbreaks, and forest fires were the most significant of these disturbances. Disturbances created conditions favorable not only to the regeneration of individual species reproduction, but also helped to create small soil mounds (tipup mounds) and other micro-environments suitable to the addition of new species to individual stands, thus helping to maintain forest diversity. In addition disturbances helped control competing vegetation for some species. On the reservation these natural disturbances are mimicked by the use of prescribed fire, chemicals, and machines.

Large-scale conflagrations are never allowed because of the risks to both people and property. However, recent scientific studies have indicated that fires played a bigger role in the life of certain forest ecosystems than has been understood in the past. Fire tends to produce a different vegetational response than can be achieved by the use of either chemicals or machinery. This has caused the reservation's foresters to review how fire has affected the forest in the past. This review has caused an evaluation of how fire can be used as an alternative to the use of chemicals and machinery. The goal is to reduce chemical use and large-scale machine disturbance.

When fire is prescribed on the reservation, the fire is managed "according to a prescribed burn plan. The plan describes the burn objectives, desired weather conditions, forest fuels, expected fire behavior, personnel and equipment assignments, and safety considerations" (Menominee Forestry Department 1995, 60–61). During the implementation of a prescribed burn plan, care is taken to make sure that all personnel and equipment called for in the plan are present on the burn site and prepared for all contingencies.

Another important tool is the amount of debris left on the ground during prescription harvests. In European forests such as the Black Forest in Germany, most debris is removed. In the Menominee Forest large amounts of debris are left to decay naturally on the forest floor.

The amount of debris differs according to the harvest prescription, but such debris helps to enrich the forest floor's topsoil and also helps to create micro-environments valuable to the maintenance of diversity. The amount of debris has to be monitored since too much can result in unexpected fires, but the branches, brush, and roots left after a prescription cut are an important element of the overall prescription.

The Silvicultural Management Processes

The management plan dictates that all silvicultural prescriptions will "apply state-of-the-art, site-specific forest management techniques to individual forest stands to meet the goal of producing the maximum quantity and quality of sawtimber while maintaining species diversity" (Menominee Forestry Department 1995, 58). Five steps are followed:

1. Before any silivicultural treatment is proposed, the stand to be treated is field surveyed. Data collected includes the stand's species composition, its timber stocking, operability from the standpoint of access and relationship to bordering stands, regeneration status, and habitat type. The field survey's goal is to determine if the stand is treatment ready. If it is ready, the survey establishes what treatments will meet management goals.

2. The silvicultural prescription is then tailored to match existing stand conditions and to effect changes necessary to achieve the desired future stand condition based on the review. The actual prescription includes a stand description, objectives, harvesting guidelines, non-timber resource management objectives and considerations, and the expected results of the treatment prescribed.

3. A silvicultural review is then initiated based on an environmental checklist which examines a variety of environmental, ecological, and cultural effects of the treatment if it is completed. This checklist is designed to verify that all environmental and cultural resources are fully analyzed. The checklist is then attached to the prescription along with a "Prescription Notice Routing Sheet." This document is

then routed to Menominee foresters, representative
foresters of the Bureau of Indian Affairs and the Wisconsin
Department of Natural Resources, and harvesting man-
agers for review and comments. If concerns are identified a
meeting is called and desirable modifications are made. A
"Compliance and Treatment Notice Routing Sheet" is then
attached to the other documents. This sheet is signed by the
Menominee Tribal Enterprise's silvicultural forester and
forest manager and the Bureau of Indian Affairs' forester.
These signatures confirm that the treatment was set up
properly and is designed to meet management goals.

4. The treatment area is then prepared for harvesting. The
   timber marking, boundary marking, and other site specific
   tasks are periodically checked by Menominee Tribal
   Enterprise personnel. When the site is fully prepared the
   BIA approves the treatment notice, both fulfilling their trust
   responsibility and providing a checks and balances system
   within the management process, and the treatment is
   started. If the BIA finds problems with the treatment at this
   juncture, a meeting between BIA and MTE foresters is held
   and a new plan is developed. The marking crew and/or the
   forester then return to the treatment site to correct
   identified problems.

5. Throughout this process a companion process is occurring.
   This process creates demonstration sites in the forest.
   Demonstration sites are used to implement new technologies and techniques on sites that can be closely monitored
   by objective means through intensive data collection and
   analysis. Research completed on these sites is peer
   reviewed before future implementations occur.

An example of demonstration sites are areas where regeneration
of white pine using the shelterwood system was first implemented. In
implementing the white pine regeneration sites, the Menominee
foresters first reviewed information from other forests around the
country, modified this information to fit into the Menominee ecosystem, and then implemented treatment. Both observation and careful
scientific data collection created demonstration site reports, and eventually the observation and objective data collected led to the current
shelterwood prescription. The result is that large sawtimber stands of

reservation white pine are being regenerated.

Some Menominee logging contractors object to prescriptions written during the last decade. The complaint is that the foresters are moving too fast, especially with shelterwood prescriptions for white pine. These loggers believe this prescription is an experiment that needs to be halted so results can be measured before the forest has too many even-aged stands. Most of the loggers acknowledge that the prescription has resulted in successful white pine regeneration, but the questions posed are, Will the seedlings grow into saplings and will the saplings reach full growth in thirty years? Or are the foresters risking leaving significant portions of the forest unproductive because of an unwise experiment?

The foresters, on the other hand, are aggressively writing shelterwood prescriptions. Data collected has shown significant regeneration success. The oldest demonstration areas now have healthy sapling stands. The foresters believe they have achieved what they set out to accomplish. Healthy communities of young white pine are again growing on the reservation. In time Menominee white pine canopies will be common again.

Forest Protection

Regeneration efforts are complemented by ongoing protection efforts. The goal of MTE's forest protection staff is "to prevent and suppress the destruction of the forest by wildfire, insects, and disease outbreaks." Most of these efforts are mitigating in character since wildfires, insect infestations, and disease outbreaks are inevitable in a forest as large as the Menominee Forest. Most mitigations are taken in concert with federal, state, and tribal agencies.

A good example is the fire protection effort. A number of multiagency agreements define responsibilities for reservation fire protection. The "Forest Land Fire Protection Agreement," signed on 4 October 1990, makes the Wisconsin Department of Natural Resources (DNR) responsible for performing all necessary fire prevention, presuppression, and suppression duties necessary to protect the forest from wildfires. The DNR's entire fire protection organization, including personnel and equipment, is available to help extinguish any fire within reservation boundaries.

A second agreement, the "Forest Land Fire Protection Cooperative Agreement," signed on 21 June 1984, provides for mutual aid in all

areas of forest fire protection between the tribe and DNR and guarantees DNR personnel trespass on Menominee lands for fire control. The
"Cooperative Agreement Between Menominee Tribal Enterprises and
Minneapolis Office, Bureau of Indian Affairs," signed 20 February
1991, allows MTE employees to assist the BIA with fire suppression
activities on the reservation and elsewhere in the United States.

Similar arrangements have been made for pest and disease control. MTE, DNR, and BIA foresters share responsibility for monitoring
pest and disease problems. When problems are identified, MTE
requests technical assistance from Wisconsin and federal specialists.
These specialists assist MTE staff in diagnosing specific forest pests
involved, identifying what tree species might be damaged, and evalu-
ating if the potential exists for a significant threat. The forest manager
designates treatment areas and directs the control effort while working with the various technical experts.

If the forest threat is significant enough, the BIA has the primary
responsibility for ensuring proper measures are taken. The BIA
forester, in this role, is authorized to request financial assistance from
the federal government to cover the costs of extensive control treatments. The BIA forester is also responsible for preparing environmental impact documents and working with tribal officials when pest control methods involve restricted-use pesticides applied using Federal
Insect and Disease Control funds.

All protection measures are developed and implemented by interagency efforts. The Bureau of Indian Affairs has primary responsibility for such measures in concert with Menominee Tribal Enterprises
and is responsible for funding needed that is greater than the MTE
budget. The Wisconsin DNR also has an important role.

This interagency focus has several effects on forest operations.
First, it increases resources MTE can use to help solve serious problems; second, it puts MTE foresters in constant contact with a broad
range of the forest management practiced in Wisconsin and on other
Indian reservations; and third, it creates a cross-fertilization of ideas
and practices that benefits each agency. In historical times this interagency aspect of forest management resulted in problems and personality conflicts. Today this is rarely a problem.

Harvesting

The harvest, or prescription application, incorporates the free market
into the forest management effort. The major goal of harvest adminis-

tration is "to work with private logging contractors to implement pre-scribed silvicultural objectives and deliver forest products in the safest, most efficient and cost-effective manner." This implementation must minimize "the damage to the residual forest" (Menominee Forestry Department 1995, 63). Both Indian and non-Indian contractors bid for work in the forest. This bidding procedure has, historically, delivered low cost, efficient harvesting for MTE.

The first step in the harvest process is accomplished by the timber harvest administration staff of MTE. This staff delineates and identifies harvest units marked for harvest by the silvicultural staff. Harvest units are designed to maximize the harvesting work's efficiency and economy. Units may or may not coincide exactly with a stand's boundaries. In addition to delineating harvest units, the timber harvest administration staff also establishes "restrictions and limitations on equipment size, seasonal timing [for the harvest], and product manufacturing and environmental concerns" for each unit. Contractors agree to these restrictions and limitations before they are awarded a contract. They are given a detailed description of the harvest unit and restrictions and limitations prior to bid deadlines (Menominee Forestry Department 1995, 63).

In the processes used to allocate harvests, Menominee contractors who hire Menominee Tribal members have the first opportunity to select harvest units if they have the necessary logging capabilities. Non-Indian contractors are allowed to take unit harvests Menominee contractors cannot handle because of contractual responsibilities or the lack of specialized equipment.

Forest roads and skid trails are developed according to strict guidelines coordinated by the harvest administration staff. Again, road and skid designs are to "limit damage to the residual forest and maintain as much of the forest acreage in timber production as possible." Roads and skids are also designed to keep logging costs down, to increase MTE profits, and to increase logging contractor profits. Current roads and skids are used whenever possible in order to limit forest damage. Proposed roads or skids are built only after a careful review and if no other option is available if a prescription's goals are to be met. Surplus roads and skid trails are always restored to timber production.

A policy is also in place that puts a priority on new harvest technologies useful to either protecting the forest environment or increasing harvest efficiencies. The harvest administration staff has the responsibility for exploring these new technologies and ensuring they meet overall forest management goals and objectives. To implement this particular policy, continuous contact between MTE and researchers who specialize in harvesting technology are maintained throughout the year.

A handbook, the MTE *Logging Practice Guide*, outlines MTE's harvest policies and is part of the effort to monitor ongoing harvests. Each contractor is responsible for following all the policies outlined in the guide while meeting the prescribed treatment's goals in the contracted time period. If a contractor fails to follow policies or to meet contracted deadlines or to minimize any negative forest effects, the contractor is eliminated from a list of eligible contractors, and the work is assigned to another contractor. Harvest monitoring has several different aspects. These include:

1. PRODUCTION LEVEL MONITORING: The harvest administration staff is responsible for making sure that scheduled harvest units are treated in a timely fashion. Foresters produce monthly reports to the MTE Forestry Committee on the status of completion for each logging contractor and projections for the future. When contractors fail to meet production quotas, the forest staff identifies the problem and recommends a solution to the MTE Forestry Committee. Past board policy has been to revoke the contract for the harvest unit and make it available for other contractors who qualify for additional work.

2. HARVEST ACCOUNTING: Accounting for timber harvested is the responsibility of the harvest administration staff. The staff verifies that: The contractor is paid the correct price for producing and delivering forest products, products are delivered to the proper mill, and harvest volumes are properly charged to the allowable cut.

3. HARVEST SAFETY: The harvest administration staff works closely with the harvest safety officer in increasing woods operation safety awareness. Staff attends all safety training seminars developed by the officer so problems in the woods can be quickly identified and immediate corrections can be made.

4. PRODUCT DELIVERY COORDINATION: Menominee Tribal Enterprises maintains ownership of all products harvested from the forest and develops delivery contracts with outside mills. Harvest administration staff is responsible for developing delivery schedules for all logging contractors that make sure mill contract delivery quo-

tas are met. In order to equalize costs incurred in delivery to these mills, volumes are evenly distributed between logging contractors.

5. PRODUCT Utilization: Harvest administration staff develops utilization standards for all harvested forest products. Sawlogs are the primary product produced during harvesting and are to be cut to maximize quality and volume. Prior to harvesting, all contractors are advised of mill specifications and are responsible for manufacturing products to their highest potential.

6. WASTE AND DAMAGE MONITORING: Harvest administration staff is responsible for monitoring and reducing harvest unit waste and damage. Due to the nature of logging, some logging damage will occur or wood will be wasted during product manufacturing. But steps are taken so that such problems do not become excessive and that all trees designated for harvest are actually cut. The forestry staff monitors each contractor during the active cutting period and does not approve a harvest completion until the prescription has been carried out properly and all waste products have been removed.

   Upon harvest completion, the staff conducts a final residue survey. This survey samples the harvest unit for actual waste, damage, and residual stocking. The calculated data is used by the inventory staff to charge waste to the annual allowable cut and summarize waste and damage by the contractor. If a contractor is found to be excessively damaging the forest or wasting potential products, this is identified and reported for action.

7. TIMBER THEFT AND TRESPASS: Timber theft and trespass is never tolerated. The forestry staff is responsible for making sure that timber theft from active harvests is not occurring.

   Timber trespass occurs where a contractor cuts beyond the boundaries of the contracted harvest or where unauthorized individuals cut standing live timber anywhere in the forest. Harvest administration works closely with the Menominee Conservation Department to make sure that cutting is only occurring where contracted. If a logging contractor is involved, MTE's policy is to immediately revoke all contracts awarded to that contractor. In any case, the incident is turned over to the proper authorities for prosecution and restitution.

(Menominee Forestry Department 1995, 65–66)

The harvest, as is illustrated by these policy summations, is as intensely managed as the silvicultural prescription process. Checks and balances are built into both processes and are rigorously followed in practice. A continuous effort is also made to upgrade both the science and technology of silvicultural and harvest practices. If you want to manage a resource in a way that lasts forever, your processes have to be sophisticated. Change and chance have to be built into the process. An example of how MTE builds change is the demonstration sites. These sites test new aspects of silvicultural prescription. The chance example comes from the conscious effort to build forest microenvironments in order to encourage the wind and other unplanned events to allow the sprouting of new species not common within an individual stand.

The harvesting effort also builds the free market into the management system. In this case contractors are closely regulated and monitored. The forest's needs are considered more important than temporary economic realities, but at the same time the free market is used to help increase harvest efficiencies while lowering harvest cost. Policies emphasize that contractor profits are important. An effort is made to assure profit for contractors while at the same time attempting to increase MTE profits. Profits and regulated free market policies enter the management plan when sawtimber moves from the forest into either MTE mills or other mills purchasing forest products.

The Integrated Forest/Mill Operation

Most of the forest's sawtimber is processed at MTE's Neopit sawmill. Some sawtimber is sold and shipped to other mills, but for over eighty years the Neopit Mill has been central to MTE operations.

Mill production machinery includes a full two-sided carriage band sawmill with kilns, rolling stock, and surface equipment. The mill's mobile equipment includes five forty-five-foot trailers. Overall, MTE has newer and more technically advanced equipment than the majority of regional sawmills (Menominee Sustainable Development Institute 1996, 43).

The mill operates throughout the year with the exception of one two-week shut-down period for repairs. On the average the mill processes between ten and eleven million board feet per year into cut, rough, or finished lumber. Some specialty products are produced according to customer specifications. The mill uses everything from the forest that enters its yards, including log waste and bark. If the bark cannot be sold, it becomes boiler fuel to heat the dry kiln. Sawdust is also used as a boiler fuel. Log waste is chopped up in a "hog," a huge machine that shreds wood waste, and is sold to pulp mills.

Logs reach the mill by logging trucks that haul sawtimber from the forest. When a truck brings in a load of logs, the logs are scaled for volume and then sorted by grade and species. Low-quality pulp logs and veneer logs are sold at this point in the process. Sawlogs go on and are loaded into the debarking area where, depending on diameter, they are run through one of the Mill's two debarkers. The scale is again recorded on the debarked logs.

> Subsequently, the logs are delivered into the mill on a bull-chain conveyer. Logs are processed down to a cant with two separate carriage band saws with computer assisted headrigs (using thin kerf technology). Lower grade lumber is sent to the resaw. Lumber is sent to the edger, where it is edged for width, and finally trimmed for length. The grader inspects the lumber, after which it is scaled and sorted by length or grade per customer's order. It may be sold green at this point or sent to the dry kiln. (Menominee Sustainable Development Institute 1996, 44)

The sawmill operation is in need of improvements currently and, from a sustainable development standpoint, needs to develop "value added products" that can increase the economic value gained from the forest, but in spite of these concerns, the Neopit mill is a facility with a distinguished history.

The mill's most important aspect, however, is that it is operated based on what is best for the forest. Financial concerns are addressed at several points in this chapter, but in the end the amount and type of timber harvested depends on what the forest can sustain. What is good for MTE, or what market demands are at any one point, never affects forest management decisions.

Each year foresters provide to the mill's staff a list of the type and amount of timber that will be harvested. The mill's marketing department then has to come up with a strategy to sell what will be coming from the forest. If hemlock is not in demand by the market, but the time has come to harvest a large volume of hemlock, then hemlock, and not trees with a higher market value, is harvested. "The forest drives the mill. The mill does not drive the forest" (Menominee Sustainable Development Institute 1996, 40).

This characteristic of the mill's operations has resulted in years when profits were low or losses were recorded. It has also hurt the effort to keep the mill as technologically advanced as possible. Investment capital has been difficult to come by. Still, MTE has a loyal customer base that initiates procurements with enough lead time to allow individual needs to be met. Most customers have bought mill products for an extended period and are willing to deal with the vagaries of a system built on sustainability in order to secure a high quality of products (Menominee Sustainable Development Institute 1996, 41).

Challenges to Sustainability

Although the overall success of Menominee Forest management is impressive, a number of challenges must still be overcome if the Menominee are going to continue sustaining their forest resource. The most serious challenges rise from changes the tribe is continuing to adjust to as an Indian people. Other challenges come from outside the reservation's borders. No ecosystem exists in isolation from other ecosystems. In the same way the Menominee people interrelate in a multitude of ways with American and world societies.

The Forest Management Plan describes what the tribe believes is at the heart of the Menominee sustainability ethic:

> Unlike many non-Indian concepts of man and nature, the Menominee people do not view themselves as separate from the forest, or the forest and its creatures independent from them. The Menominee culture exists in harmony with Mother Earth, understanding the circle of life. The forest, properly treated, will sustain the tribe with economic, cultural, and spiritual values for today and for future generations. This has been taught and practiced on Menominee lands for more than eight thousand years, and accounts for the quality of the forest on the reservation today. (Menominee Forestry Department, 1995 83)

A constant reservation question concerns whether or not this particular ethic can survive in each succeeding generation. More Menominee are alive today than at any one point in history. Much of the tribal membership resides off reservation in places like Chicago, Milwaukee, and Minneapolis. The values of Menominee children born in these places are different from those on the reservation. As the tribe's non-reservation population grows, can the values that have sustained the forest for so long be maintained? How can long-held values be transmitted from one generation to the next, especially in the inner city where unhealthy values are the norm and are sometimes even necessary if the teenager is to survive in that environment?

Making these questions even more difficult is the sustained management program's success. As the years have passed the forest has increased in value. "The Menominee Forest today is worth several hundred million dollars in timber value alone." The Menominee are still not a wealthy people. Those living in America's inner cities, and many reservation families, are among the United States' poorest people. Forest liquidation could financially benefit all living Menominee families. In the past Menominee values have kept the forest sacrosanct from such concerns, but will those values continue to be as strong as they are today?

This particular question raises yet another question. The mill operation is facing increasingly difficult questions. Due to increased machinery costs, new technology, and an increasingly global market economy, reinvestment in the mill and forest operations is essential. Still, reinvestment costs are increasing, not decreasing, as competitive technologies surface from around the world. Profit levels in the timber business are being shrunk by such technologies and the ability of large lumber companies to increase harvest volumes. Because the tribe's long-term ethic calls for forest protection before economic needs are addressed, the forest and its land cannot be used as collateral to generate bank loans. This means the tribe can generate reinvestment dollars only by increasing timber cut volumes. This creates a dilemma. On one hand MTE needs to reinvest in order to maintain a place in the market. On the other hand MTE cannot reduce forestland in order to generate profits needed to secure reinvestment dollars.

Making this dilemma even worse is the current tightening of federal and state budgets. Since many reservation people survive on welfare programs, what is going to happen when welfare dollars are reduced and those without jobs cannot find jobs? As was seen in the economic chapter of this book, the number of jobs available on the reservation, although increasing, falls far short of those needed to employ every Menominee who needs work. As the management plan points out, "if the Menominee Tribe was forced to support tribal communities as a result of federal, or even state, governmental actions such as funding cuts or increased taxation, the pressure [on MTE] to make money for the short-term needs would be intense" (Menominee Forestry Department 1995, 83).

These cultural and economic concerns are only a part of the sustainability challenge. Another major concern is forest fragmentation. Forest fragmentation occurs when land is changed from forest to permanent nonforestry housing, recreation, or industrial use. A special danger occurs when the forest, instead of maintaining contiguity, is broken up and one stand no longer blends into other stands. Fragmentation changes the forest environment and negatively impacts its ecosystems.

The fact that small amounts of land have been eroded from the Menominee Forest and put to other uses in recent years makes forest fragmentation a long-term possibility. This is not a serious problem today, but the foresters take this threat seriously and continually bring it to the tribal membership's attention.

Another major concern is political. The foresters have long believed that interagency relationships are a key component of sustainability. The sharing of professional expertise has not only allowed the Menominee to continue improving management techniques and technologies, but has also allowed foresters to seek advice and recommendations from outside agencies. This advice has often strengthened the forestry staff's ability to resolve threats to the forest. The concern is that a reservation politics might develop which discourages open door policies and emphasizes Menominee solutions instead of collaborative solutions to forest problems. As the management plan emphasizes, "if outside cooperation is weakened or lost, the Menominee tribe would risk losing a valuable source of technical assistance, and, worse, restrict communities and agencies outside the reservation from sharing the Menominee land ethic." To the foresters, tribal isolation could lead to future disaster.

Green Cross Certification and Green Marketing

One of the most ambitious projects underway at MTE currently is the effort to maintain Green Cross certification by the Scientific Certification System and Smart Wood certification from the Rain Forest Alliance. These certifications assure consumers that wood products are from a sustainably managed forest. This effort is part of the attempt to address the threats to sustainability outlined above. As noted earlier, one of MTE's enduring challenges has been to maintain a strong market when sustainability demands that the forest, and not the law of supply and demand, dictate its produce. Larry Waukau, MTE's president, is convinced a green market will form around issues raised by sustainable development, and he is hoping that Green Cross and Smart Wood certifications will help MTE tap that market, increasing profit margins and helping MTE avoid losses like those recorded in the past (Waukau 1994, conversation).

Although both certifications have been achieved and are in place, the long-term effect of such certifications in the marketplace are still undetermined. MTE has been able, in specific instances, to increase the profit margin realized from individual sales as a result of certifications, but these instances have not occurred often enough to constitute a trend.

The Value-Added Project

The other major plan for addressing sustainability challenges is formally called the "Menominee Tribal Enterprises Revitalization/Value-Added Expansion Project" or the "value-added" project. This particular project has three elements:

1. Enhance existing wood processing efficiency in a way that reduces operating costs and increases throughput/grade recovery. Implement improved preventive maintenance/ safety programs. Expand the existing quality control program. Improve inventory management/production scheduling processes. Increase management and technical education. Improve material handling. Expand kiln capacity.

2. Maximize resource processing in order to efficiently process all sawlog, bolt wood, and pulp logs at MTE facilities. Install a log sorting system. Install a bolt wood processing facility. Install a whole log chipping facility. Install automated lumber sorting and stacking system with new trim line.

3. Create a value-added wood products manufacturing capability designed to move MTE into producing downstream semi-finished

and finished wood products. The capability would include: Fingerjointing, cutstock, sanding, finishing, flooring, edge gluing, molding, panel/lay-up/press, and veneer slicing.

(Menominee Sustainable Development Institute 1996, 46)

None of these efforts ensure long-term sustainability, but, at the same time, they are part of the flexible system evolved to maintain the forest. New sustainable policies and practices will have to be achieved if the long-term Menominee goal of ecosystem sustainability is to be achieved. Still, current management policies and procedures are part of the bedrock that will lead to the evolution of these future policies and practices.

9

An Analysis of the Menominee
Model of Sustainable Development

The Interrelationship of Human
and Nature's Cycles

IN CHAPTER 2 eight aspects of the Menominee experience were
identified as crucial to understanding the Menominee model of sustainable development. The historic, legal, spiritual, cultural, ethical,
political, technological/scientific, and economic aspects of
Menominee life were described as being "interrelated in complex
ways that often defy precise description." They were also called time
streams. Human cultures and environments change. These changes
cause other changes to occur, thus making management of sustainable
development processes complex and difficult. Any management system created to address such complexities and difficulties must have, if
success is to be realized, flexibility that does not compromise
immutable sustainable principles.

This book indicates that Menominee sustainability arises from two
primary, interrelated sources. At the heart of Menominee sustainability are the cultural, spiritual, and ethical beliefs of the Menominee people. Historical events, however, have shaped and changed these
beliefs. When the Menominee first came into contact with the
Frenchman Jean Nicolet on Green Bay's shores, they were clearly living sustainably on the earth. As William Ruckelshaus noted in an article entitled "Toward a Sustainable World,"

> Sustainability was the original economy of our species. Preindustrial
> peoples lived sustainably because they had to: if they did not, if they

expanded their populations beyond the available resource base, then
sooner or later they starved or had to migrate. The sustainability of
their way of life was maintained by a particular consciousness
regarding nature: the people were spiritually connected to the animals and plants on which they subsisted; they were part of the landscape, or of nature, not set apart as masters. (Ruckleshaus 1989, 112)

As the Menominee became integrated into the European, then the
American, economy, changes occurred in Menominee cultural, spiritual, and ethical beliefs. The original "economy of sustainability," over
time, became a set of cultural and ethical beliefs fashioned by
Menominee reactions to the economic and assimilative forces that
became increasingly powerful in their lives. These beliefs became
powerful values arising from the spirituality that has characterized the
Menominee from pre-European times to today.

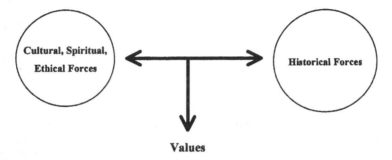

Figure 9-1. Illustration of How Menominee Values Have Formed

Menominee Values

In values research conducted on the reservation in the early 1990s
(Cokain 1993), the researchers used the coined word *ecority* to characterize one primary way the Menominee approach the world. They
defined ecority as "the capacity, skills and personal, organizational, or
conceptual influence to enable persons to take authority for the world
"to enhance its beauty and balance through creative technology in
ways that have worldwide influence" (Hall et al. 1991, 179). This particular concept, along with other Menominee values, at least partially
explains the origins of the Menominee sustainable ethic. According to
the value theory posited by Hall et al., four levels of consciousness

exist. Three factors determine these levels: the perception of the world by the individual, the perception by the individual about how they are functioning within the world they perceive, and "what needs the self seeks to satisfy" (Hall et al. 1991, 8).

Levels, or phases, of consciousness described by Hall et al., are: 1) the world as mystery/the self as center, 2) the world as a problem/the need for belonging, 3) the world as project and invention, and 4) the world as mystery-cared-for/self-as-lifegiver. Individuals tend to move through these levels as they age, although individuals can become permanently "stuck" at any one level.

Values are the internal forces which motivate and provide criteria for shaping our lives as human beings. They arise from the "priorities" that reflect an individual's worldview. Worldviews are shaped by the consciousness level individuals are living within at any one point in time. Values develop meaning as the result of internal images that precede the streams of thought, feeling, and action that make up individual lives. As individuals, we react to the external world by transforming these images "into action through language." Language is a motivating force at the heart of who individuals, institutions, and people are (Hall et al. 1991, 22).

During the first phase of consciousness human values are related to safety and survival needs. Values formed out of this phase promote skills development that enable children to meet basic physical and psychological needs. In the second phase, which usually begins at the age of eleven according to Jean Piaget, the Swiss psychologist, humans begin the struggle to become part of a significant human community. Individuals attempt to become accepted by others as well as to have their work recognized. Acceptance and recognition allow them to feel a sense of belonging. During this phase values such as competence, confidence, and self esteem become critical (Piaget 1968). In the third phase humans move from dependence upon other human beings toward independence. Physiological needs can largely be met by the individual, so they are free to develop values such as human dignity, service, and vocation. The last phase constitutes the highest phase. In this phase the concept of interdependence comes to the forefront of conscious thought, and values like harmony, interdependent relations, and synergy become critical (Hall et al. 1991, 8–20).

As values form out of each consciousness level, their formation occurs in stages. Stage A in each of the phases relates to how the "Self" views its basic needs as it reacts to what it sees and experiences in the environment. Stage B in each phase describes how the "Self" acts to manipulate the environment as it seeks to satisfy the individual's basic needs. An inner, creative response in Stage B helps the individual to "cope, mold and recreate the world" so that it becomes more fulfilling and satisfying (Hall, et al. 1991, 23).

According to this particular value theory, tension exists between stages. The value of self-preservation in Stage A demands that we protect ourselves from physical harm in a threatening and hostile world. In Stage B this worldview drives our search for security in our lives. We learn, in our search for security, to build shelters and protections of various kinds for our families, institutions, and organizations (Hall et. al. 1991, 23).

Stage A values are called "Primary Values" and stage B are "Means Values" in the matrix Hall, et al., use to graph the value-sets of different groups of people. Administration of the Hall-Tonna Values Index, based upon the "125 value words that have been identified in our written and spoken language that consistently appear throughout the lifespan of individuals and organizations," established that Menominee tend to hold more fourth phase, or the world as mystery-cared-for/self-as-lifegiver, values than other American communities studied by the same researchers. In most American communities most individuals administered the Hall-Tonna Value Index held third phase values. Two years of research on the Reservation identified the following characteristic Menominee value sets:

Table 9-1
Menominee Values

Primary Values	
Self Worth	Work/Labor
Wonder/Awe/Fate	Truth/Wisdom, Integrated
Insight	
Ecority/Aesthetics	Equality/Liberation
Faith/Risk/Vision	Family/Belonging
Integration/Wholeness	Ritual/Communication
Security	Service/Vocation
Self-Competence/	Art/Beauty as Pure Value
Conficence	

Means Values

Territory/Security	Support/Peer
Workmanship/Art/Craft	Wonder/Curiosity/Nature
Unity/Diversity	Tradition
Economics/Success	Education/Knowledge/Insight
Efficiency/Planning	Endurance/Patience
Equity/Rights	Friendship/Belonging
Health/Healing/Harmony	Honor
Interdependence	Management
Membership/Institution	Mission/Objectives
Mutuality	Responsibility/Accountability
Ownership	Productivity
Responsibility	Sharing/Listening/Trust
Simplicity/Play	Social Affirmation
Adaptability/Flexibility	Achievement/Success
Disconcernment/Communal	Courtesy/Hospitality
Corporation/New Order	Cooperation/Complementarity
Community/Personalist	Care/Nature

(Cokain 1993)

Such values as ecority/aesthetics, truth/wisdom, integrated insight, integration/wholeness, and beauty-as-pure-value all belong to the fourth phase of consciousness. The word ecority in this list is especially important to this book's analysis of sustainable development. Ecority drives individuals to take responsibility for the world. It also gives them the desire to enhance the world's beauty and its balance through a creative use of technology. The truth/wisdom/integrated insight value set leads individuals to an intense pursuit of ultimate truths. Integration/wholeness leads to an individual's effort "to organize the personality (mind and body) into a coordinated whole" (Hall, et al. 1991, 181).

This discussion does not validate a theory of human values. Research conducted by Hall, Taylor, Kalven, Rosen, and Cokain does not explain the complex layering and integrating human development processes described by scholars like Jean Piaget, Erik Erikson, and Abraham Maslow. Still, this research shows that the Menominee approach individual and community responsibilities in ways different from those common in other American communities. It further associ-

ates this difference with the concepts of ecority, integration, whole-
ness, balance, aesthetics, truth, and wisdom—all ideas central to the
pre-European Menominee culture. It also provides a framework for
understanding how such clusters of values, shaped by historical and
cultural/spiritual/ethical forces, have resulted in the Menominee
approach to the political, technological/scientific, and economic
aspects of life.

From Values to a Society Based on Sustainable Development

Menominee values have had a number of effects upon political/insti-
tutional, technological/scientific, and economic aspects of Menominee
life. They have led to the Menominee conviction that they are a single
people. This conviction is held in spite of the fact that the Menominee
have intermarried with other races and peoples since the earliest
French fur traders came to their territory. It reinforces Menominee
values, culture, spirituality, and ethics since all of these help the Tribe
define itself as a people apart from the larger society.

This sense of unity, however, is counterbalanced by the Lobster
Bucket Democracy of Menominee institutions and politics. An
extraordinary number of political and bureaucratic institutions struggle for predominance on the reservation. Political and bureaucratic
divisions are exacerbated in turn by economic, religious, and acculturative divisions that make Menominee political life complex and
intense. Coalitions form and disappear based upon family, friend, and
ally relationships in an endless political dance that brings one coalition
into power just in time for another coalition to begin undermining the
political legitimacy of those with office, position, and, patronage.

Within this political structure new ideas are constantly brought
forward, but changes are incremental. The idea of change has little
currency on the Reservation, except in the sense of a reservation politi-
cian calling for a leadership change. The lobster bucket keeps the society in constant motion, pulling down even good leaders before they
can amass enough power to actually effect change. Both financial and
political success, in this context, creates suspicion and distrust.
Attempts to change what is, unless the ideas driving the change have
become part of the Menominee cultural fabric, come up against an
invisible barrier that refuses to move until a consensus of Menominee
has been achieved. Once consensus has been achieved, the change

becomes radical rather than incremental. At this point the change fought over so intensely becomes a part of the overall Menominee cultural awareness, and tribal culture is adjusted to fit the new paradigm.

Those championing the new paradigm are quickly put back into the Lobster bucket after their victory, however. Competing institutions start developing agendas based on the new paradigm. Issues are quickly re-formed in the new social and cultural context. Before long a major political figure such as Ada Deer, the Menominee leader most responsible for Restoration, even though the policies she fought for became a paradigm and triumphant, has her legitimacy threatened and then falls from power. If such a leader chooses to stay on the reservation and "lead the good" fight on reservation terms, he or she has a chance to return to power once the legitimacy of the newly victorious leaders is undermined. But no leader escapes Shirley Daly's lobster bucket.

Menominee economics are caught within the limitations imposed by the Lobster Bucket. On the one hand a significant disparity exists between the wealthier Menominee, who belong solidly in the American middle class, and the poorer Menominee, who are among the nation's most impoverished people. On the other hand one of the Lobster Bucket's rules is that all Menominee should advance up the economic scale together and not as individuals. At the same time, a modified belief in economic rationality exists. Tribal businesses engage in the free marketplace with all the effort put forward by the most dedicated corporate chief executive officers. Individual reservation businesses are valued for the increased employment and income they bring the tribe. Menominee also willingly work for profit-driven businesses off the reservation.

The tension between the lobster bucket political system and its distrust of political and economic success and the acceptance of economic rationality built into the American free market system provides part of the lobster bucket's energy. Discussions about economic success, efficiency, and planning erupt with nearly every political campaign. The sense is always given that opponents running for office are trying to feather their nest at the tribe's expense. The argument is usually posited as follows: The tribe would be doing better economically if the casino, mill, or tribal government were run with better planning and more efficiency. If those on the other side would only put tribal interests first, all Menominee people, not just the elected few, would prosper.

Menominee economics are thus tied into the lobster bucket. Progress tends to be incremental as change is resisted until consensus is reached. Then the tribe as a whole, meeting in its competing institutions, decides that the change proposed will result in a tribal good. The change happens.

A good example is the Menominee decision to go into the casino business even though serious questions were raised about the legality of such an effort. From outside the reservation this decision seemed to be made almost accidentally. A federal lawsuit involving a California tribe and the State of California was won by the tribe in the United States Supreme Court. Menominee tribal government officials were hesitant to move into the casino business, even though the tribe had been in the high stakes bingo business for a number of years and approved of the revenue gambling could generate. Tribal leaders hesitated because of the risk posed by a hostile state government position against Indian gambling. A Menominee entrepreneur then started his own casino, arguing that as a tribal member he was protected from prosecution as a result of the Supreme Court decision. The issue then dominated reservation political discussion. A consensus formed, based upon countless discussions about the good and ill of gambling that had been going on since high stakes bingo became a possibility. Tribal government then closed down the individual entrepreneur and went forward into the casino business in spite of the risk it posed. After consensus is formed change is dramatic.

One consensual economic value rarely threatened has been communal ownership of Menominee land. The Menominee agree on two major principles concerning land: 1) that all Menominee land should be held by the tribe in communal ownership and 2) that since land is not the property of individuals, individuals only have limited rights to benefit from the land's resources. Rights are given to individuals, furthermore, with the understanding that both the tribe as a whole and the individual should derive benefits from the individual's use. This use must also not violate the ethic that insists the forest and Menominee environment must exist forever. These limitations act as brakes on individual initiative, allowing the Menominee to avoid the Tragedy of the Commons.

The forest and its management are also largely held above political battles. One of the core cultural values is ecority. This means the forest must be managed for profit and beauty by creatively using technology to achieve those ends. No sane person would sell Mother Earth, but every sane person would acknowledge that the Mother is responsible for taking care of her children. The forest and Menominee ecosystems should be used to the maximum advantage by those entrusted with such responsibilities. The earth is also a spiritual, holy place and as such must be treated with the care and attention a holy place deserves. This means the forest and its ecosystems should be managed in ways that strengthen and preserve the natural order while still serving the community's human needs.

All of these elements have become a part of the Menominee heritage. Derived from history, this heritage is part of the tribe's culture and spirituality. "It is said of the Menominee people that the sacredness of the land is their very body, the values of the culture are their very soul, and the water is their very blood" (Pecore 1992, 12). According to the Menominee ethic, the land, the forest, the Menominee ecosystem, and the Menominee themselves are interrelated with bonds that should not be broken. All are sacrosanct and must be preserved forever.

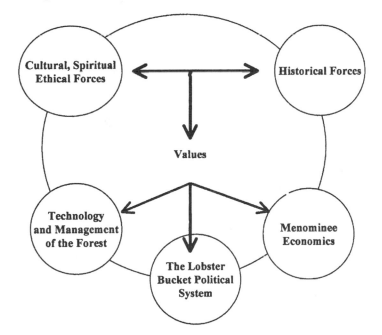

Figure 9-2. Illustration of the Interaction of Forces within Menominee Sustainable Development.

Menominee sustainability and development, then, have been created from the interaction of cultural and historical forces that have resulted in sets of values held by the Menominee people. These values, in an interaction with cultural and historical forces, have, in turn, created the lobster bucket political system as well as the complex Menominee economic system and the consensual forest management system based upon aesthetic and creative technology values. All of these values and systems interacting together allow consensual change and adaptation to forces imposed by the world at large.

This complex circle of historical and cultural forces, values, technology, management, politics, institutions, economics, and ethics then creates Menominee sustainable development. It relates human actions and desires to the healthy functioning of ecosystems, thus opposing unhealthy economic exploitation of resources while pushing forward the ideal that all people should benefit from the earth's produce both as groups and individuals.

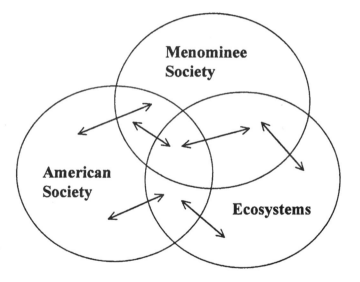

Figure 9-3. Illustration of the Interaction between Menominee Society, American Society, and Ecosystems.

The Menominee model of sustainability creates a circle of interactions that results in a concrete set of Menominee values, but the Menominee circle is affected and shares by boundary areas with both the larger human universe as well as with the various natural systems that are so much a Menominee concern. All of these elements interact with and change each other, creating the pools and whirlpools of interaction that were called "time-streams" in chapter 2.

Within the boundary areas of interaction change is the most immediate and powerful. This is where what is happening within American society affects one or more elements of Menominee society, thus changing it. American society can also impact the Menominee ecosystem through a causality chain. A distant ecosystem can be changed. This change can then force changes in neighboring ecosystems, the changes eventually impacting the Menominee ecosystem. The reservation, the forest, and its ecosystems are not closed systems. They cannot evolve without influences from the interactions that occur within the complexity of their relationships. They are open systems.

The Menominee, as the result of values that see the living and nonliving nonhuman communities that share Menominee land as important and worthy of protection as the Menominee themselves, create a rational ecology. This form of rationality establishes the limits of environmental exploitation and also defines the acceptability of economic, social, political, and legal rationalities that affect the tribe and its larger ecological community. This rational ecology is the circle that binds other rationalities and interrelationships into a coherent, working system, the glue that helps define the Menominee as a people. Its best expression is in the management plans and activities of those who manage and harvest the Menominee Forest, but it is also a part of all the other elements that have made the Menominee who they are.

Key Sustainable Development Characteristics

The complex interactions of the elements that have created Menominee sustainable development are only part of the story, however. Each element also has characteristics that are part of the way both sustainability and efforts at development work on the reservation. These characteristics are as important to note as the way each of the elements interact toward what the ecological economist Herman Daly has described as a steady state economy, an economy where "the size and stock and the rate" of man's economic activities are not "so large relative to the total environment that they obstruct the natural ecological processes," through pollution or other negative environmental effects, "which form the biophysical foundations of wealth" (Daly 1980, 20). Summary statements of these characteristics are as follows:

1. The loss of the pre-European Menominee land base, combined with the corresponding destruction of forests on
   lands the Menominee once inhabited, has made the
   Menominee especially aware of the value of their remaining lands, forest, and ecosystems. This awareness has provided long-term underpinning to efforts made at achieving
   a sustainable use of resources.

2. The Menominee associate their well-being as individuals
   with the maintenance of the land community's health. This
   is partially an association that rises from Menominee history as the result of the tribe's suffering when it was
   removed from its ancestral lands and is partially a cul-
   tural/spiritual characteristic. Menominee culture and spirituality teaches that the Menominee, as human beings, rose
   from the Earth out of the form of a great bear. The
   Menominee came from the Earth. They had ancestors that
   were like the bear, wolf, moose, eagle, and crane. They are
   also the children of Mother Earth. In the traditional family
   relationship, the mother is to be honored, but she also has
   the responsibility to nurture her children. This means that
   while the children must take care of the Mother, the Mother
   must also provide for her children through the provision of
   the land's produce (natural resources).

3. A great spiritual diversity exists on the reservation, ranging
   from traditional beliefs to Catholicism to Mormonism to
   the Native American Church. Open acknowledgment of
   spirituality is a constant of reservation life. Constant attention to public prayer stresses, irrespective of religious institutional membership, the importance of human humility in
   relation to the Great Spirit's universal creations.

4. Songs, dancing, and other celebrations are also constant in
   Menominee life. This keeps Menominee in touch with their
   history and cultural/spiritual beliefs. These events help to
   give a sense to all Menominee of who they are as a people.
   They also help to remind children and adults of what their
   relationship to the Great Spirit and the earth should be.

5. The acculturation process the Menominee have gone
   through has created a complex, modern society that
   embraces the free market world of the Americans while, at
   the same time, balancing free market beliefs with the spiri-

tual/cultural constants in Menominee life. Conflicts inherent in the complex groupings of traditional people versus more contemporary Menominee provides part of the reservation's lobster bucket political system's energy. Menominee democracy has led to the creation of multiple competing reservation institutions which have, in turn, led to decision-making that allows only the most incremental changes to occur in the way the reservation operates and Menominee live. Menominee incrementalism is more grudging of change than the incremental American political system with which it coexists and interacts. What this means to sustainability is that once sustainability has become a part of the consensual Menominee paradigm, it is a powerful force not easily modified. The same is true for the development side of the sustainable development equation. Change is difficult to effect. Economic development, although widely recognized as a tribal good, is, nevertheless, a change agent widely mistrusted. Therefore, although the lobster bucket political system is constantly generating new economic development ideas and efforts are continually moving forward on economic development, unless change is coming from timber harvesting and manufacturing or gaming where a consensus already exists, not much succeeds. Change is thwarted simply because the legitimacy of reservation political groups, along with their agendas, is overcome by the intense maneuvering on the reservation.

6. When consensus is built around a potential economic, cultural, or spiritual change, however, the system acts like a snapping rubber band. Change comes quickly, and it forces the democratic institutions to adjust to the new consensual reality. This is especially true for development efforts as the examples of Restoration and the development of casino gambling attest.

7. Consensual changes occur only within a specific framework. This framework comes from cultural constants such as the communal ownership of tribal lands and the values and characteristics described in points 1–4 above.

8. The Menominee have a strong sense of intergenerational equity. This sense of equity not only encompasses the need

to protect the interests of future generations but also stresses the importance of Menominee ancestors to tribal identity. Only if you honor those who have come before can you truly protect the interests of those who are yet to be. It is this continuum, interacting with the continuum of the earth's past and future, that makes sustainability and development within the limits of sustainability more important than current economic realities.

9. The Menominee believe that if the forest is destroyed, then the culture, the very body, soul, and blood, of the Menominee people will also be destroyed. The bond between humans and the earth's ecosystems must never be broken. The power of this idea comes, at least partially, from the sense of the importance of past and future as they relate to the continuum of Menominee ancestors to unborn generations.

10. In this context the Menominee believe they have the responsibility for making sure their forest is preserved forever in a healthy, diverse state.

11. In order to preserve the forest forever, the Menominee believe that the most advanced science and technology must be applied. MTE and the Forestry Department are continually exploring new technologies that can help achieve the best stewardship possible. To a large degree the Menominee Forest's history is based upon how scientific and technological advances have been applied to forest and ecosystem management

12. A related characteristic has been the Menominee willingness to use American jurisprudence as a form of technology in their efforts to protect their forest heritage. Several important law suits have played key roles in ensuring that the Menominee continued to control the resources they retained through the Treaty of the Wolf River. The tribe has also been aggressive in pursuing legislative remedies to complaints when legal remedies have fallen short of achieving desired goals. An example of this particular willingness is the Menominee Restoration Act which effectively stopped reservation land sales, based on the Legend Lake development, and helped the tribe preserve most of its land holdings.

13. The Menominee have also been skilled, from the treaty period of their history onward, at giving in to the acculturative forces of American society and then complaining, legislating, and suing away the changes they did not want in the first place. One reason the Menominee have been able to maintain their presence on aboriginal Wisconsin territory is this tribal characteristic. These tactics have also been effective in helping to establish the legal framework for long-term reservation sustainability.

14. The Menominee have created an extreme version of democracy, Lobster Bucket Democracy. To outsiders this political system often seems irrational and unnecessarily mean-spirited. However, at least in the Menominee case, it protects the key cultural, spiritual, sustainable, and unifying aspects of the Menominee people. What appears on the surface to be an engine of tribal instability is, instead, a protection device that helps to maintain consensual decision-making. New ideas and proposals are a constant part of this particular political system as different political and institutional groups attempt to form a consensus that will make their policy effect on Menominee history dominant in Menominee life. Even in rare cases where consensus forms, however, an individual leader's legitimacy is still attacked. Such attacks, sooner or later, lead to the leader's downfall.

15. Reservation political discourse always emphasizes the good of the Menominee people. This constant reminder of the primary importance of the tribe and not the individual has powerful effects on how reservation economic development efforts are pursued. Such discourse is also a constant reminder about the importance of tribal consensual values.

16. Part of the Lobster Bucket Democracy is the idea that every Menominee should have a voice in Tribal operations. Institutions like the General Council are designed to achieve this ideal. However, a plethora of reservation institutions play a role in giving the maximum number of Menominee a political voice. These institutions form another part of the energy that drives Lobster Bucket Democracy. They also help fuel the questions about political legitimacy of individual leaders or groups of leaders

that eventually lead to that leadership's downfall. These competing institutions provide the ideas and policies that drive tribal government operations and jostle together toward consensual decisions.

17. The combination of Lobster Bucket Democracy and consensual decision-making give the Menominee a tremendous ability to hold outside assimilative forces at bay. The Menominee are continually reaffirming who they are through a process of discussion and conflict. Discussion allows an airing-out of commonalties and differences while the conflict allows decisions about exactly what the tribe believes. One tribal institution may support termination and champion its cause, causing turmoil and conflict in political circles, but the challenge inherent in that institution's efforts causes the forces of reaffirmation of consensual values to reassert themselves. Nonconsensual efforts at change fail because of the strength of consensual forces in reservation life. This is an extremely complex tribal characteristic, but an understanding of it is essential to any understanding of Menominee cultural strength.

18. To date, poverty, in spite of the development of the gaming industry on the reservation, is a powerful, complex force. It has not ended.

19. The Menominee participate fully, in a number of different ways, in the free market economy. They are as eager to make a profit and participate in the American consumer society as any other group living in the United States. In recent years the reservation's small entrepreneurial class has been growing.

20. However, the Menominee reject the transformative economy which sees value in land only as it relates to increased individual wealth. Menominee own private property, and some Menominee own land, but the primary reservation consensus rejects individual land ownership in favor of communal land ownership. It is this communal ownership, combined with Menominee spiritual, cultural, and historical values, that allows the tribe to avoid the tragedy of the commons. The Menominee have always believed that the ownership of personal property by individuals was accept-

able. This sense of the importance of personal property allows the Menominee to participate fully in the free market. Still, those who are too successful raise suspicion, and too much success undercuts an individual's legitimacy within the tribe. This causes a balance between the individual's drive for success and the tribe's overall sense that all tribal members should succeed equally.

21. While fully supporting economic rationality, modified by rejection of the idea of land as a product that can benefit individuals versus the entire tribe, the tribe is willing to forego the benefits of full rationality by lessening productivity in order to gain employment and other social benefits. This is especially evident in Neopit sawmill operations where jobs have taken precedence over productivity enhancement for decades. Still, a current move to improve productivity within the context of maintaining current employment levels is underway. The idea is to develop value-added processes as well as productivity efficiencies to mill operations in order to achieve a greater financial return while maintaining social benefits.

22. The forest and Menominee ecosystems are managed based upon silvicultural prescriptions based upon the application of forest science to the Menominee environment. A management intensity practiced by the Menominee Forestry Department and Menominee Tribal Enterprises is comprehensive in its approaches and prescriptions. Experiments are continuously undertaken in small demonstration stands but are slowly implemented. These actions help the Menominee work toward preserving the forest forever.

23. The Menominee see "green" marketing and certification efforts as key in their efforts to maintain sustainability as a value accruing to future generations.

All of these tribal characteristics help form the substance of the Menominee model of sustainable development.

10

Conclusion

The basic message to be gained from the Menominee model of sustainable development would seem to be straightforward. If humankind decides to take a long-range view toward managing earth's resources, then neither scientific, technological, nor economic development has to stop. Short-term sacrifices might be necessary, but in the long term human ingenuity combined with stewardship can avoid the worst pitfalls foreseen by analysts like William Ophuls and Stephen Boyan, Jr., Paul and Anne Ehrlich, or Lester W. Milbrath.

By believing their forest has to be preserved forever and accepting the economic sacrifices a long-term viewpoint has imposed on the tribe, the Menominee have put themselves in an enviable position. Although reservation poverty still exists, it is lessening, partially because of gaming, and the future looks bright. At the same time the Menominee Forest, in its splendor, is healthier and more productive than at any other point in history, dating back to when the Menominee first canoed up the Wolf River to their new reservation home.

In this message the forest and Menominee people are a powerful metaphor for what is possible for the planet. The forest is the earth. The Menominee are all of mankind. By deciding to preserve the Earth forever and living with sacrifices inherent in this ideal and wisely using whatever tools are made available by science, technology, culture, and history, the earth can be preserved and made more productive and beautiful than at any other time in human history. By trying to grow and sustain a garden that will last forever, using a Christian metaphor, we can make the garden bloom. By developing these metaphors, a mantra can be constructed:

> We can throw away the throw-away society and tend the
> garden of the forest until our hearts are lost in the
> wonder of paths through the woods.

We can sing our children's children's song into the distance
     of the far future and bring alive profits good for our
     hearts and spirits.
We can pay attention to the details of what our forest will
     become, and we can bring alive promises hidden in
     the richness of soils.
We can become like shadows beneath the canopies of the
     great white pine, trilliums blooming white in the early
     spring, songs of birds inside thickets of young birch,
     the weaving sound of wind whispering into the fire-
     leaves of red maple, the soft running of a bobcat
     beside the river-music of the Wolf,
And in our becoming, we can come to know the richness of
     who we are, the glory of our long-lived cities, the rela-
     tionships of our spirits to the spirits of sheep, bear,
     trout, beaver, crane, eagle, rock, river, oak, and moun-
     tain.
We can reach for stars while still bending to the earth and lis-
     tening to the deep sound of its breathing—
Its breath, our breath, the sound of time passing into the sun-
     rise of tomorrow.

By simply changing perspective, the Menominee model says, we
can have a healthy, productive ecosphere while still striving toward
prosperity, a strong culture, and good, safe lives. By accepting the discipline a long-range view imposes, we can have nature's glories while
nurturing our physical, mental, and spiritual well-being as humankind.

Unfortunately, neither mantras, metaphors, nor conceptions of
Shangri-la tell us how to get from where we are to the promise inherent in the Menominee Forest. The true test of the Menominee model's
usefulness lies in its ability or inability to help chart a future path that
results in preservation rather than simply using and discarding. The
list of challenges to the earth and its environment is formidable; an
examination of a single topic that is part of any single environmental
challenge can take a scientist a lifetime of study. How can we successfully identify and remediate non-point-source pollution?

Part of what is accomplished by examining the Menominee model
is an explanation of how the Menominee act differently from mainstream society and why their decision-making has evolved in the way
that it has. The rejection of the transformative economy, the heart of

modern economic systems, alone makes the Menominee unusual. These are days when the world is becoming increasingly capitalistic and personal-property oriented. The wise-use movement with its assertion that property ownership is an inalienable right as important as those inalienable rights set forth in the *Declaration of Independence* provides evidence of just how deep the sense of property ownership goes in American society.

The Menominee do not stop with their repudiation of the American sense of property rights. They go much further in their analysis of the human relationship to the earth. The Menominee rejection of the idea that every decision must be based upon the nation's economic good is another example of where they are at odds with a society that elects its President largely upon the economy's relative strength or weakness during times leading up to an election. The common good, extended to animal and plant kingdoms, is a familiar theme both in Menominee cosmology and thought, taking the idea one step beyond that championed in Christian teachings.

Understanding just how different a perspective the Menominee live can be secured from an analysis of *The Wisdom of the Spotted Owl* by Steven Lewis Yaffee. In his book Yaffee describes how the Pacific Northwest forests and ecosystems were managed before the spotted owl controversy erupted in 1989:

> [T]imber and dependent community interests, horrified by the potential impacts of the court injunctions on timber supply and regional economies, 'appealed' the judicial decisions [which attempted to protect the spotted owl's habitat in the Pacific Northwest's remaining old growth forests] to the U.S. Congress, an arena in which they had strong supporters. (Yaffee 1994, xvi)

Forest and ecosystem management decisions, according to Yaffee, were largely made upon a set of traditional behaviors which included:

- An overwhelming adherence to timber production as the primary organizational objective of the Forest Service and the Bureau of Land Management in western Oregon and Washington;

- A tendency to view resources other than timber, such as wildlife or recreation, as either secondary or adjuncts to the timber management program;

- A proclivity to find solutions to difficult choices through elaborate technical analyses and planning processes whether they were warranted or not; and

- The development of a FS [Forest Service] organizational image and style as a tightly controlled "Can Do" agency that ironically made change in direction more difficult.

(Yaffee 1994, 3)

The Forest Service and Bureau of Land Management have been changing perspectives in recent years. A greater appreciation of the various publics that demand different uses for Forest Service managed lands dominates current thinking; but the Menominee have never viewed production as the primary value in organizing forest management. Management, instead, has been built from the effort to achieve long-term sustainability.

The Menominee also have a close affinity to the non-timber aspects of their forest land, valuing noncommercial wildlife and water resources as highly as timber resources. This book describes some of the sources of those values, thus illustrating their strength in Menominee culture. The Menominee, through their institutions, have made extremely difficult choices, accepting continued poverty instead of destroying their forest's long-term viability, and, as the chapter on Lobster Bucket Democracy shows, the Can-Do spirit is neither the organizational image nor style of reservation decision-making.

American society is full of structures alien to the way Menominee react to life. The idea from political science of iron triangle political structures where private industry, political committees in Congress or state government, and federal and state bureaucracies join together to protect the economic interests of a single industry, such as the logging industry, is a case in point.

Michael E. Kraft describes how dominant iron triangles have been in U.S. natural resource policy. He first points out that they were created as the result of the way the West was settled. In the 1800s the federal government "sought to encourage rapid development of its vast holdings in the West" by adopting "public policies toward that end." Then "the unfettered and heavily subsidized use of public lands for such a long period created the belief among the recipients" [of the benefits from iron triangles] that they had a right to receive such benefits indefinitely."

> The legacy affects natural resource policy even today. For years, pow-
> erful western constituencies have dominated these natural resource
> policies, forming projectionist subgovernments or sub-systems in
> association with members of key congressional committees and exec-
> utive agencies. Such alliances operate autonomously with little polit-
> ical visibility. Conflict is minimal and is easily resolved through
> logrolling, or mutually beneficial bargaining in which each party
> may gain its goals. (Kraft 1996, 67–68)

In Menominee culture the political strengths of competing institutions would make quick work of any iron triangle that would try to
form on the reservation—especially if the triangle's purpose was to
favor the economic interests of one group over another. The triangle's
leaders would quickly find themselves in the lobster bucket's bottom.

The predominance of the idea that economic freedom and political freedom are linked together by the forge of American democracy is
equally out of step with the way the Menominee live and react. The
free marketplace has a place in the Menominee universe. They have no
problem at all competing with other lumber producers and forcing
free market disciplines on loggers and suppliers. But the idea that
individuals should be unconstrained by either cultural norms or government regulation is not acceptable. Access to the forest and other
elements of the Menominee environment are strictly controlled and
restricted. Menominee culture as it exists today acts as a governor on
Menominee individuality. This book has cited numerous examples
of Menominee ideas and practices that modify classical economic
rationality and the practice of individual free will in favor of long-term
Menominee survivability.

This book has given a host of clues about how different the
Menominee are from the rest of American society. This brief discussion
only touches the surface of these differences. But a core question is,
With these differences, do the Menominee really have substance to
add to the sustainability discussion? Or is the example they provide so
removed from the modern world and what that world is becoming
that it has little relevance?

These are fair questions. A gulf exists between the attitude
expressed by Milton and Rose Friedman in their book *Free to Choose*
(Friedman and Friedman 1979) and the Menominee approach to economy, politics, and culture. To the Friedmans,

> A society that puts equality—in the sense of equality of outcome—
> ahead of freedom will end up with neither equality nor freedom. The

use of force to achieve quality will destroy freedom, and the force, introduced for good purposes, will end up in the hands of people who use it to promote their own interest.

On the other hand, a society that puts freedom first will, as a happy by-product, end up with both greater freedom and greater equality. (Friedman and Friedman 1979, 148)

The Menominee do not equate equality with the necessity to enforce the will of one political group over another. They believe, instead, in the primacy of community over individual free will while still allowing complete freedom of expression. The Friedmans' belief that cultural demands for equality inevitably lead to dictatorship is fallacious. Still, the Friedmans express what the American majority believes. The United States is the free world's leader. Freedom finds its expression in the individual, not in the community.

In spite of such cultural positions, however, a cause for hope exists. Menominee sustainability arises from values shaped by Menominee history, culture, and spirituality. This complex of forces have created a politics and an approach to economics, science, and technology that supports sustainable development practices. These practices and the human and environmental relationships created have resulted in a sustainable community.

Neither countries like the United States nor individual communities within a country are likely to reach the same sustainable solution the Menominee have reached. The Menominee value of ecority, which is at the heart of Menominee sustainability, is a creative value. The Menominee value the universe as a whole in all of its splendor, and they use technologies ranging from those that originated far in their past to those derived from contemporary forest science in an attempt to improve the world they know. The Menominee are a micro-example useful for deriving macro concepts about sustainable development.

Inherent in the Menominee model is a sense that their way is not the only way to preserve the earth and improve its peoples' lot. Menominee culture, and American Indian culture in general, is as much about survival and sustaining core cultural values as about anything else. The essence of what the Menominee model says is that individual places and cultures should, out of their own experiences, fashion creative technological, cultural, spiritual, and human solutions to challenges posed by the need for long-term preservation.

More important than a critical path to sustainability are the sources of sustainable effort. Have the people making up a community interpreted their history in such a way as to understand the importance of long-term preservation of resources and the ecosystem? Does the community's culture help its citizenry understand their relationship with the earth, its creatures, and the ecosphere? Does the community have a spirituality that allows it to develop strong relationships between the biota and human culture? Does an understanding and appreciation of generational continuity within the community's culture exist? Is the community balanced in its approach to questions of economics, culture, and spirituality? Does it understand that human well-being and security are related to cultural, spiritual, and environmental well-being? Does the search for knowledge and technology have a context based upon the long-term welfare of human non-human generations of life that share this planet? Are political mechanisms designed to force review and reflection upon actions taken that impact the community? Are individual freedoms defined in ways that protect both the earth and human community? Does the community possess a strong sense of the value of community? Does the community have a strong sense of itself?

This list of questions does not exhaust questions important as sustainable indicators. One lesson to be drawn from the Menominee model is that too close a definition of any one set of indicators may miss the point. The Menominee have changed their approach to sustainability over time in reaction to the multitude of circumstances and historical and other forces that have impacted tribal life. The tribe's flexibility, grounded in the idea that the forest's and Menominee well-being are inextricably linked, have allowed Menominee sustainability to survive time's test.

Reasons for pessimism concerning the achievement of a healthy and sustainable world culture are numerous. Some of these were described at the beginning of this chapter, and many more exist. A litany of environmental problems outlines with powerful force the reasons for pessimism.

The case for optimism comes from another direction—in the cultural and intellectual legacy provided by both contemporary and historical writers, scientists, economists, and philosophers. This legacy constructed a foundation upon which the environmental movement could be built. It also serves as the legacy upon which a sustainable contemporary culture can be founded. The Menominee have given a gift to humankind. They have proven that sustainability over extended periods of time is possible. They have provided a model that others can study and, hopefully, apply to different situations in different communities in different ways. But, a question remains: What are the traditions, ethics, and spiritualities that can lead a wealthy country like the United States toward becoming a sustainable community?

Thoreau, Jefferson, and Powell

Three visions that could help shape a new sustainably developed world can be found in the work of Henry David Thoreau, Thomas Jefferson, and John Wesley Powell. Significant differences exist between the sustainable world their visions might lead to and the one the Menominee have actually created, but at its core the Menominee model allows events, personalities, and values to change while the goal of sustaining the forest forever remains the same.

The work of these three scientists and philosophers differs considerably. They do not present a cohesive vision that puts before us a critical sustainable path any more than the Menominee present such a path. Still, taken together, their work forms a foundation upon which sustainability can be built. They are important American historical figures with a powerful mythos, and this can help transform the tradition they create into a tradition for sustainability.

One of the most attractive sustainable development ideas ever developed, and certainly one of the simplest, came from Thoreau. In his famous book *Walden* he noted that:

> [W]e live meanly, like ants; though the fable tells us that we were long ago changed into men; like pygmies we fight with cranes; it is error upon error, and clout upon clout, and our best virtue has for its occasion a superfluous and evitable wretchedness. Our life is frittered away by detail. An honest man has hardly need to count more than his ten fingers, or in extreme cases he may add his ten toes, and lump the rest. Simplicity, simplicity, simplicity! I say, let your affairs be as two or three, and not a hundred or a thousand; instead of a million count half a dozen, and keep your accounts on your thumb-nail. In the midst of this chopping sea of civilized life, such are the clouds and storms and quicksands and thousand-and-one items to be allowed for, that a man has to live, if he would not founder and got to the bottom and not make his port at all, by dead reckoning, and he must be a great calculator indeed who succeeds. Simplify. Simplify. Instead of three meals a day, if it be necessary, eat but one; instead of a hundred dishes, five; and reduce other things in proportion. (Thoreau 1946, 73)

Thoreau believed, like Ralph Waldo Emerson, that humans should be self-reliant. He meant, by moving to his cabin by Walden Pond, to set an example. He planted his bean field, worked the earth that had before only grown "cinquefoil, blackberries, johnswort, and the like" (Thoreau 1946, 133), made sure his income from this labor exceeded his "outgoes" and lived a life of simplicity, independence, and thought. He preached the values of wilderness and Nature.

What he wanted most was to discover reality, to "explore the private sea, the Atlantic and Pacific Ocean of one's being" (Thoreau 1946, 282). He said that "the universe is wider than our views of it" (Thoreau 1946, 281). He thought a miracle had occurred when a bug hatched inside the apple-wood of a kitchen table that had sat in a farmer's dining room for sixty years.

Thoreau's work differs significantly from the Menominee sense of sustainability. It is much too self-centered and individualistic in tone to have a Menominee origin, but it still has qualities that have made it an environmental classic. Thoreau's self-reliance was based upon an individual search for a spiritual sense of reality. The individual was not to live meanly "like ants" in an ant-hive. They were to notice life in all its forms, whether the form was waterbugs swimming in Walden Pond in the spring or johnswort. They were to glory in existence—their own as well as that of the Natural World, and "simplify, simplify, simplify" in order to discover the universe of which they were a part.

What has made Thoreau's work so powerful is its realization of the importance of Nature to the individual's search for self-truth. Only by placing himself by Walden Pond and growing beans and gathering berries for sustenance can mankind begin to understand where they are placed and why. Individual man must be self-reliant, an independent spirit marching to his own drummer, but he must also realize his place and seek truth from that place, realizing that in a speck of dust the imagination can connect itself, by following the paths put before us by Nature, to the fires of stars.

A good counterpoint to Thoreau's emphasis on individual self-reliance is Thomas Jefferson's effort to think through the process of nation-building. Before the build-up to the War of 1812, Jefferson believed the new nation he had helped create should primarily be an agricultural nation. As he said in *Notes on Virginia,*

> The political economists of Europe have established it as a principle that every state should endeavor to manufacture for itself: and this principle, like many others, we transfer to America, without calcu-

lating the difference of circumstance which should often produce a different result. In Europe the lands are either cultivated, or locked up against the cultivator. Manufacture must therefore be resorted to of necessity not of choice, to support the surplus of their people. But we have an immensity of land courting the industry of the husband-man. Is it best then that all our citizens should be employed in its improvement, or that one half should be called off from that to exercise manufactures and handicraft arts for the other. Those who labour in the earth are the chosen people of God, if ever he had a chosen people, whose breasts he had made his peculiar deposit for substantial and genuine virtue. (Jefferson 1984, 290)

With the build-up to the War of 1812, however, he changed his mind. While still considering those who worked the land the chosen few of God, the war, with its embargo and port blockages, had him note in a letter to P. S. Dupont de Nemours that "the spirit of manufacture has taken deep root among us, and its foundations are laid at too great expense to be abandoned." He continued by saying that:

[T]his mass of *household* manufacture, unseen by the public eye, and so much greater than what is seen, is such at present, that let our intercourse with England be opened when it may, not one half the amount of what we have heretofore taken from her will ever again be demanded. The great call from the country has hitherto been of coarse goods. These are now made in our families, and the advantage is too sensible to ever be relinquished, (Jefferson 1984, 1209)

Jefferson's emphasis upon the small farm combined with family-based, or small commercial, manufacturing was part of his belief that the bonds that tied men to their country and liberty were not those of economics and property, but were the bonds that came, in Wendell Berry's words, from "the investment in a place and a community of work, devotion, knowledge, memory and association" (Berry 1977, 144).

Like the Menominee, Jefferson believed science and education should work toward improvement of the earth's produce. Science had the virtue of making human progress possible, and education

[E]ngrafts a new man on the native stock, and improves what in his nature was vicious and perverse into qualities of virtue and social worth. And it cannot be but that each generation succeeding to the knowledge acquired by all those who preceded it, adding to it their own acquisitions and discoveries, and handing the mass down for successive and constant accumulation, must advance the knowledge and well-being of mankind, not infinitely, as some have said, but indefinitely, and to a term which no one can fix and foresee. (Jefferson 1984, 461)

Spirituality and its connection with the earth was also an integral part of Jefferson's overall approach to political economy even when he argued that freedom of religion was one of the key foundations for a just society.

Jefferson proposed to govern this nation of small farmers, shop-keepers, and manufacturers by dividing "every county into hundreds, of such size that all the children of each will be within reach of a central school in it." This division,

> [B]esides a school, should have a justice of the peace, a constable and a captain of militia. Those officers, or some others within the hundred, should be a corporation to manage all its concerns, to take care of its roads, its poor, and its police by patrols, &c, (as the select men of the Eastern townships.) Every hundred should elect one or two jurors to serve where requisite, and all other elections should be made in the hundreds separately, and the votes of all the hundreds be brought together. . . . These little republics would be the main strength of the great one. (Jefferson 1984, 1226–27)

Although Jefferson did say in a letter that he agreed with John Adams "that there is a natural aristocracy among men" (Jefferson 1984, 1305), he supported the democracy of "the select men of the Eastern townships" that stood before the citizens in meeting and discussed the day's important issues.

Jefferson's concept of democracy was not as radical as that of the Menominee. It did not claim that small groups and institutions should be formed in a small geographic area for the purpose of developing competing visions and then going to political war until consensus could be reached. Nor did he have quite the same idea that the Menominee and John Wesley Powell share about the necessity of conforming human habitation and labor to the environment. But he did champion as close a human relationship to the land as possible, foretelling the Walden Pond experiences of Thoreau, claiming that those that worked the land were the closest of humankind to God.

From a Menominee standpoint the ideas of John Wesley Powell, although probably less attractive than those of either Thoreau or Jefferson to contemporary Americans, are in many ways more promising. Wallace Stegner, the Pulitzer Prize-winning fiction writer, in his Powell biography provides a good description of the reason for Powell's proposals. At the time, Powell had not yet founded the Bureau of Ethnology that produced the work of Walter Hoffman on the Menomini. He was the nation's most famous explorer who had opened much of the arid West through dangerous rafting expeditions down the Green and Colorado Rivers. He had also, by the time of his *Report on the Arid Region of the United States,* founded the U.S. Geological Survey and become a fixture in the political maneuvering of federal agencies in Washington, D.C. As Stegner states:

> Almost alone among his contemporaries he [Powell] looked at the Arid Region and saw neither desert nor garden. What he saw was the single compelling unity that the region possessed: except in local islanded areas its rainfall was less than twenty inches a year, and twenty inches he took, with slight modifications for the peculiarly concentrated rainfall of the Dakotas, to be the minimum needed to support agriculture without irrigation. In no other part of the United States did that aridity pertain, though in what he at first called the subarid zone between about the 97th and 100th meridians there were sure to be dry years. (Stegner 1954, 223–24)

Considering the possibilities for settlement of the arid West, Powell first recognized that drought would always be a constant threat. This fact made water the region's controlling wealth. The first principle he proposed for settlement was that "The growth and prosperity of the Arid Region will depend largely upon a land system which will comply with the requirements of the conditions" of the region (Powell 1962, 37). He believed, in other words, that lands should be parceled out according to their uses. Farms working irrigated lands could exist on 80 acres, but for pasture farms he recommended 2,560 acres, or four full sections. He also said that

> [T]he pasturage farms need small bodies of irrigable land; the division of these lands should be controlled by topographic features to give water fronts; residences of the pasturage lands should be grouped; the pasturage farms cannot be fenced—they must be occupied in common. (Powell 1962, 40)

Even though he believed individuals should own the land where they earn their material well-being, he also wanted the land to be organized efficiently, developing communities around water sources that could then work the pasturage of larger tracts radiating outward from the community center.

Refining his ideas of community, in his consideration of how to organize the effort to irrigate lands close enough to a water source for irrigation ditches to be built, Powell suggested what amounted to irrigation cooperatives. These were to be called irrigation districts. Such districts were to consist of "any nine or more persons who may be entitled to acquire a homestead from the public lands," and these districts, based upon the commonality of water need and usage, would provide the basis for Western communities (Powell 1962, 42).

As Stegner noted in his analysis of Powell's ideas, these proposals were

> revolutionary. It was as bold as Powell's plunge down the canyoned river, for it challenged not only the initiative, individualism, and competitiveness which were quite as marked as co-operation in the American character . . . but it challenged as well the folklore bred through generations of frontier farmers in a country of plentiful rain. It challenged too the men who were already beginning to ride like robber barons and kings over the public domain, and the corporations who were already, with Scottish and English and American capital, beginning to acquire those water-bearing half- and quarter-sections upon whose possession depended the control of range to support a cattle empire. (Stegner 1954, 229–30)

Much of what Powell has to say fits into the Menominee model. His notion that human settlement should be based upon what we would call today the land's environmental conditions fits the Menominee environmental ethic. Powell's insistence, too, that science be used a tool to understand the land and its needs so that scientific principles could govern human land use patterns would mirror similar attitudes in the approach the Menominee take toward forest management. His ideas about the necessity of human cooperation and community also fit into the Menominee model. Clearly, a culture based partially upon Powell's ideas, driven by strong reactions against the history of environmental degradation, would seem to provide an ecologically rational basis for a sustainable society.

His ideas are not quite the same as Menominee ideas. He believed in the sanctity of the free market system and the importance of the small farmer as opposed to the cattle barons. He did not attempt to link culture and the environment as the Menominee do, but his ideas could help form a cultural basis for an ecological rationality nevertheless.

Taken together, the work of these eminent Americans suggests that an American sustainability would have a more individual core than that held by the Menominee. Community would be centered in the uses of the land, responding to the land's needs as well as human needs. Democracy would inform this society, providing the intense consideration of proposals to development so necessary to a society based upon long-term versus short-term needs. Also important is the concern for political jurisdictions that allow a selectman-type democracy where those involved in governing are constrained and empowered by friends and neighbors instead of faceless masses and special interests.

A Contemporary View of Sustainability

Most contemporary contributions to a heritage for sustainability start with the work of Aldo Leopold and then refract into the work of hundreds of other observers. In the "The Land Ethic," as pointed out in chapter 6, Leopold enlarged "the boundaries of the community to include soils, waters, plants, or, collectively: the land." He asked for human responsibility toward this community, holding that humankind's image as Nature's conqueror is self-defeating. A human is a "plain member and citizen" of the land community, he declared, and since "man-made" community changes are different than those caused by evolution, humankind is responsible for achieving harmony "between men and land."

He then went on to argue against a "system of conservation based solely on economic self interest." He extolled the value of a songbird, or of Draba, "the smallest flower that blows." He saw humanity as a "plain" citizen of the biota with the responsibility to protect the complex interactions that make the biota healthy. As a functioning part of the ecosystem, humankind had to behave responsibly to the land community, subjugating individual profit for the sake of community good.

In one remarkable essay, "Thinking Like a Mountain," he went beyond his lifelong attempt to define human relationships to the land and embraced the value of a mountain and a wolf as others that have a powerful life beyond human purview. He first describes coming upon an old wolf and her grown pups "joined in a welcoming melee of wagging tails and playful maulings." Then he describes his reaction to finding the pack:

> In those days we had never heard of passing up a chance to kill a wolf. In a second we were pumping lead into the pack, but with more

excitement than accuracy: how to aim a steep downhill shot is always confusing. When our rifles were empty, the old wolf was down, and a pup was dragging a leg into impassable slide-rocks.

We reached the old wolf in time to watch a fierce green fire dying in her eyes. I realized then, and have known ever since, that there was something new to me in those eyes—something known only to her and the mountain. (Leopold 1966, 137–38)

In this essay Leopold suggests that not only must we protect the land community of which we are a part, but we must also recognize that part of the community exists outside our knowledge. Not only must we protect what we see and understand, but we must also protect and preserve the community's more difficult part, that which exists in the green fire in the eyes of a dying wolf in communion with its mountain home.

These, and other, Leopold ideas refract, with increasing richness, through much of modern environmental and sustainable development thought. The economist E. F. Schumacher argues against the short-term view of economic self-interest that Leopold stood against when he states:

From an economic point of view, the central concept of wisdom is permanence. We must study the economics of permanence. Nothing makes economic sense unless its continuance for a long time can be projected without running into absurdities. There can be "growth" towards a limited objective, but there cannot be unlimited, generalized growth. It is more than likely, as Gandhi said, that "Earth provides enough to satisfy every man's need, but not for every man's greed." Permanence is incompatible with a predatory attitude which rejoices in the fact that "what were luxuries for our fathers have become necessities for us." (Schumacher 1975, 33)

The land and its community's economic interests are different from the human economy described by classical and neoclassical economists. Permanence, not growth, is the standard that measures ecosystem stability.

The lessons Talbot Page, the biologist, teaches, have their source both in Leopold's ideas about the land community and his sense of otherness in the dying fire of the wolf's green eyes. To Page, the first lesson humans must learn is that "there is no sharp boundary between humans and other animals." Therefore we should extend "traditional moral concepts, such as duty, utility, rights, autonomy, and justice," and apply them to both animals and humans.

Lesson two is that human defining boundaries are fading. According to Page, biology is proving that human beings, like apes and chimpanzees, emerge from "social interaction." Society is not created out of individual actions and ideas. This means that we need to stop approaching problems such as intergenerational equity, population size, and species extinction with "highly individuated tools." We need to approach these problems by discovering a value theory "which does not rely on the predefined (exogenous) individual."

Lesson three is that ecology is replacing economics as the dismal science. Ecological problems identified to date seem to be endless and include: "species extinction, soil depletion, forest loss, mineral depletion, carbon dioxide and CFC buildup, pollution, and so on." This lesson "challenges the optimistic view that a sustainable balance between depletion and renewal would be struck automatically by market forces, without explicit policy attention." Furthermore, the third lesson leads to two major challenges: To "develop a framework in which the possible policy actions (large or small) can be placed," and "to develop a system of valuation for the possible policy actions which does not foundationally rely on the predefined individual."

The last lesson is "that the appropriate time scale for understanding resource and population stability is much longer than we are used to." This has to be compared to the market's "myopic" processes which often act in response to even shorter term events such as elections and political pressures. "The challenge to a value theory, from this lesson, is to find ways of valuing the distant future, perhaps over three centuries or more" (Page 1991, 58–65).

Talbot's work goes far beyond what is contained in Leopold's writing. Still, like most of Leopold's intellectual heirs, Talbot owes much to his teaching. The idea that human ethics and morality need to be applied to both the land community we know and the community we do not know are core Leopold thoughts. Talbot's idea that no sharp boundary exists between humans and animals is only an extension of Leopold's thought.

The biggest criticism of Leopold's work has been that it is too focused on community and not enough on the individual. To a number of biologists and ethicists, individual pain and suffering, human or animal, is of paramount importance. These commentators have argued that Leopold's emphasis on the community's whole misses the point that individuals and community are interlinked. Only by meeting individual needs can the community as a whole be served. Baird Callicott answers this criticism by pointing out that it misses a theme about individuals found in Leopold's writing. Ecosystems are important to Leopold, but humans, as individuals, are part of the system and plain members of the community in the same way individual animals and plants are members.

The Richness of the Sustainable Heritage

The richness of the sustainable heritage since Leopold has been proliferating at an ever faster rate. From the work of deep ecologists such as Bill Sessions, George Devall, Gary Snyder, and Arne Naess, through the economics of ecological economists like Herman Daly and Robert Costanza, to the biology of Edward O. Wilson and James Lovelock, to the ecology of Barry Commoner and Kenneth Boulding, to the environmental philosophy of Baird Callicott and John Passmore, to essayists like Wendell Berry and Edward Abbey, intellectual, spiritual, and cultural foundations for a sustainable society have been placed.

A classic work by Herman E. Daly and John B. Cobb Jr., *For the Common Good*, presents just one vision of how to achieve a sustainable world. Going back to Aristotle, they describe the meaning of two ancient words, *chrematistics*, "the branch of political economy relating to the manipulation of property and wealth so as to maximize short-term monetary exchange value to the owner," and *oikonomia*, "the management of the household over the long run." Then they ask that modern humankind expand the scope of *oikonomia* "to include the larger community of the land, of shared values, resources, biomes, institutions, language, and history." In this context they argue against the short-term destruction of "*chrematistics* and for *oikonomia*, a community economics (Daly and Cobb, 1989, pp. 138-139).

This vision, like Lovelock's Gaia vision, Thomas Berry's *The Dream of the Earth*, the Biophilia hypothesis of Edward O. Wilson, or the work done at the 1992 Rio de Janeiro Earth Summit, all contain ideas that can lead toward a sustainable world. Hope can be derived from the Menominee model because the intellectual, cultural, historical, and spiritual framework for a sustainable world is present in western culture.

The Problem with the Heritage

The problem with people, however, is that we are inherently short-sighted. Almost all humankind looks to the future, but that future is short. It usually encompasses our lifetime, and sometimes our children's and grandchildren's lifetime, and maybe, sometimes, even a generation or two preceding us. But in Menominee terms this is just a blip in the forest's timeline. The forest is forever; the Menominee people are a forest people, bound into the forest's continued existence. Time spans considered by most Americans do not lend themselves comfortably to the longer-term vision of Menominee sustainability.[1] More importantly, awareness of short time spans will not create a world system of sustainability. Unless we can get out of Boulding's "spaceship earth," we need to be careful not to foul our nest.

The Menominee have not altered the way individuals think. They are motivated by individual incentives the way other Americans are. They eagerly buy the newest technology, struggle for political power and predominance, benefit family and friends when they have the power to do so, and try to make the best of their individual lives. The problems the Menominee community faces can be appalling. The problems of any city ghetto can be found in one of the most beautiful environments on earth. Drug use, gangs, incest, murder, drunkenness, gambling addiction, fraud, suicide, and other human sorrows and pain can be found in the midst of towering white pines and white rapids thundering over granite outcroppings. Menominee problems can be, and have been, overstated. Most Menominee are good people who live good lives based upon an admirable citizenship. The Menominee are people like other people.

What makes Menominee different are the tribal structures evolved to handle how individuals are allowed to act. Maddening, unjustified, hateful, and unacceptable to most observers, the political lobster bucket is a powerful reservation force. When individuals are pulled down from the bucket's rim when they climb too high, and they have to deal with the pulling-down process's emotions, the lobster bucket system can seem to be an unmitigated evil.

But, in the end, the lobster bucket creates norms by which Menominee society lives. Power is not invested in any individual's or family's hands. Individuals have to acknowledge the power of Menominee culture as well as to show an appreciation for Menominee generations that have graced ancestral lands for greater than five thousand years. The reservation and forest is home. Therefore, the reservation and forest must be preserved and cherished.

A Sustainable Cultural Message

Jefferson, Thoreau, Powell, Leopold, Daly and Cobb, and other authors, scientists, and philosophers mentioned in this book all instill aspects of a sustainable cultural message to American society. But our institutions have not developed in a way to encode the beliefs, just beneath the surface of American thoughts and actions, into actions that recognize the sanctity of humans as individual members and citizens of the land community.

What the Menominee microcosm teaches above all other lessons is an institutional lesson. Time is a great changer. Transformation is in the order of eons. Fish become land animals which become birds, apes, and bears, and bears, in Menominee cosmology, become Menominee. In the human community some aspects of human institutions must achieve long-term rigidity, or else the earth and its creatures shall be destroyed.

In the Menominee context forest, land, and culture are key to the tribe's institutional persona. Neither the forest, land, nor culture can be sacrificed. These keystones have allowed the Menominee to survive Winnebago, Huron, French, British, and American incursions. These keystones, buffeted by outside forces, have evolved the Menominee institutions of sustainability.

What the world needs to focus on is clear if we are to sustain ourselves upon spaceship earth. We need to evolve and maintain a set of institutions, grounded by cultural and spiritual touchstones, able to adapt over time while ensuring that *some* rules of individual behavior are maintained in an immutable state. That is, we can manage our forest in different ways at different times, but we cannot sacrifice it. We are caretakers for all the generations of "the larger community of the land, of shared values, resources, biomes, institutions, language, and history" beyond.

We are, in Gary Snyder's language, in need of an ancient awareness:

> The archaic religion is to kill god and eat him. Or her. The shimmering food-chain, the food-web, is the scary, beautiful condition of the

biosphere. Subsistence people live without excuses. The blood is on your own hands as you divide the liver from the gallbladder. You have watched the color fade on the glimmer of the trout. A subsistence economy is a sacramental economy because it has faced up to one of the critical problems of life and death: the taking of life for food. (Snyder 1990, 184)

We need to recapture again an intermixing of human culture, the Other, and a sacramental economy (the saying of grace at the dinner table) if Gaia and life dependent upon her abundance are not to be destroyed.

The Menominee model provides an outline of one possible set of institutions that can create and maintain sustainability. Jefferson, Thoreau, Powell, Leopold, and others provide elements for outlines of other sets of institutions. The point is that institutions have to have a core immutability: The environment is sacrosanct. We must understand that when we kill the least of the environment's parts, we are out of grace. We have killed ourselves.

Among these core values are a more egalitarian society, an abiding sense of ecority, a willingness to be caretakers of generational time spans and the land community, and a valuing of individual community members while disciplining them to the needs of unborn generations. We must function as an integrative part of the forest, the people, and the spirit—the community.

The question facing us is, Is this the impossible dream? The idealistic equivalent of Brooks Farm during the time of Henry Thoreau and Ralph Waldo Emerson? A good, glorious, hopeful dream that can never be realized? A mantra that can be said, but that cannot be translated into human actions? Or are we doomed to a catalog of catastrophes? Shortages, upheavals, and pestilences?

The good news is that humans are rational. The Menominee microcosm shows how adaptable human beings can be in a changing universe. We confront issues and concerns whether they are caused by the coming of the French or environmental disaster. These confrontations are not always successful and do not always happen quickly enough, but rationality is still part of human character.

We live in a system, the ecosphere, with limits. These limits define what is sustainable over time. The more we allow ourselves to approach these limits, the more survival problems we are likely to face as generations pass.

None of the multitudes of challenges facing us should be understated. They are all serious and sometimes seem insurmountable. The most obvious of these problems is human population. How well could the Menominee manage their forest if their population went unchecked and they could not draw from outside resources as inputs into the local economy or as a place for excess population to move? The lesson provided by Middle Village's construction off the reservation and away from the forest is clear. The Menominee have had to build on purchased land to solve current population problems and preserve the forest. They cannot preserve the forest if too many people reside in the forest.

Population problems just lead the list of intractable problems that have to be faced. Another major issue is whether we are truly willing to live with an economy that imposes limits. None of us want to do away with indoor plumbing or give up our automobiles. Those who live without those luxuries often crave them. Our economics is designed to lead to the greatest possible distribution of goods to the greatest possible number of people. Both classical and neoclassical economics work efficiently toward this overall goal.

The problem with modern economics is not that it is wrong when it is so obviously a powerful engine for much good. The problem is that it is not aimed toward long-term survival, but toward the production and consumption of goods. Are we really ready to evolve an economic system that includes this new goal of ecological rationality within the traditional sense of economic rationality?

One of the central problems brought forth by Menominee sustainability is that of the commons. The Menominee have modified economic rationality in order to strengthen their commons, their land, and their resource base. They have rejected in the process the wise-use argument that ownership of land is an inalienable right. They have instead built their economics around preservation rather than exploitation.

Achieving this particular good is relatively easy with a forest or a water system, however. Forests are alive. By working them, you can extract wealth, while leaving the forest intact. You can maintain clean water systems as long as your neighbors fail to build a mine or a factory that pollutes them. But the U.S. economy is partially built on nonrenewable resources like oil, coal, gas, and minerals. Can these resources be maintained as common resources in the same way that forests and water resources can? Is the world prepared to lessen its feeling that the ownership of private property is a sanctified right?

Can we modify our views of the transformational economy in ways that will aid in the preservation of the environment rather than continue to encourage its exploitation?

Questions like these lead to other questions related to the Menominee sense of spirituality and culture. There exists incontrovertible evidence that a strong religious revival is in process in the United States today. But do the spiritual feelings engendered by this revival lead toward what Menominee spirituality leads to—a sense of reverence for the earth and all the members of Leopold's land community? Can American culture truly become a sustainable culture, one which integrates into its unspoken beliefs the sanctity of the earth, unborn generations, and a more egalitarian human condition?

These issues, problems, and questions can easily lead toward pessimism and even fear. What do we do when the spaceship earth has been fouled beyond redemption? What do we do if we do not evolve institutions with immutable long-term survival values at their core? The Menominee model, as a microcosm, provides good service. Strong, but mutable, institutions are going to be needed to handle complex environmental issues if long-term world answers to sustainability are to be found.

The Menominee model, as a microcosm, provides a wonderful prism with which to look at the macrocosm and the challenges it faces. The model provides hints about answers to some challenges even though it does not provide all the answers. It is not the mantra that will save the earth.

Still, the Menominee people, their forest, history, culture, and spirit leading to community, are to admired. They have provided a valuable lesson to the world.

Notes

Chapter 1. Menominee Indian Tribe of Wisconsin

1. Under its current president, Larry Waukau, the Enterprise has made a profit for four of the past five years. This is the longest string of profitable years in recent memory.

Chapter 2. Menominee Sustainable Development Model

1. One Menominee informant who read this manuscript promptly informed me that there are wolves and have been wolves on the reservation all of the time the tribe has occupied Menominee lands. Since then this comment, although not scientifically verifiable, has been related to me by a number of other Menominee people.

2. Today white pine regeneration is no longer the worry it once was. An elaborate silvicultural prescription regimen has been put in place on the reservation that guarantees successful regeneration of white pine stands. There will be a significant period of time when the beloved white pine canopy of the reservation will not be as pronounced as in previous decades since the methodologies for white pine regeneration have only recently been put into place and the existing white pine canopies are aging and will have to be cut during this generation of Menominee. But new canopies will, within decades, replace the disappearing canopies.

3. The Menominee have been using the phrase *sustained yield* for a long time. Unfortunately, during recent decades sustained yield has also become a common phrase used by the logging industry as a whole. The Menominee use the term in a much different way than the rest of the industry. This has resulted in confusion about what kind of forestry the Menominee practice. Mladenhof and Pastor, discussing the sustained yield idea common in forest

management literature, said that the phrase was based on a paradigm that postulated a "regenerating state in the forest landscape" that "conveyed the impression that such a state could be maintained (sustained) to provide a constant timber supply (yield). They continued their description of the sustained yield idea by pointing out that "classic silviculture developed in this simplified landscape with a paradigm that there is a maximum potential site productivity using devices such as site index and then manipulating the stand to bring actual productivity up to potential, which can theoretically be maintained" (Mladenhof and Pastor 1933, 157). When the Menominee use the phrase, as Marshall Pecore has pointed out, their "vision" is of "a management process" that will "allow the forest to be harvested at a rate" that achieves "a perennial balance between annual growth and natural mortality and the production of timber through selective processes" (Pecore 1992, 13). This idea forces the Menominee, at least during the modern era, to concentrate on all of the systems that affect the forest. The idea is not to simply make sure that growth in the forest matches the amount of wood fiber removed, but to address the long-term health issues of the forest environment and its various ecosystems at the same time. Big timber production companies like Weyerhaeuser use the "sustained yield" phrase to describe what they are achieving when they clear cut an area and then plant a plantation of pine trees in order to replace the wood fiber removed during the clear cut. The Menominee sense of sustained yield is much more complicated. The character of the forest has to be maintained even as trees are removed from the forest. If you simply replace what you have taken, then, according to the Menominee, the forest disappears.

4. The Menominee harvested timber before 1865, but the harvests were small, primarily for firewood and other local uses. In 1865 three hundred thousand board feet were cut. In 1866 the Indian agent, who had to authorize the Menominee to make use of their land, did not allow a cut. It was not until 1885 that the cut exceeded the 1865 harvest.

Chapter 4. Culture and the Spirit

1. Deloria, an attorney, is one of the most respected Indian voices among Indian people. His words are often used to explain the Indian reaction to a variety of policy issues affecting Indian tribes in today's world. This particular quote, although taken from Deloria's book, was also used in *Freedom with Reservation*, a political tract used by the Menominee opposed to Termination to develop the case for the Menominee Restoration Act.

Chapter 5. Politics, Political Character, and Institutions

1. Economic rationality is the idea that those making economic decisions have available to them all the information necessary to make an economic decision in their best interest. Then the rational decision is always made.

2. Just how big a victory was scored by the Menominee when they

avoided the Allotment Act is outlined by Peroff. "When the Allotment Act was finally discarded in 1934, only about 35 percent of the 140 million acres held by Indians in 1887 remained under federal trust in reservation, and approximately ninety thousand Indians had become landless" (Peroff 1982, 12).

Chapter 6. Menominee Economics

1. The next two passages, constructed from Menominee County files, are composite sketches. The actual studies are not publishable due to privacy considerations. These two sketches reflect the kinds of problems faced by a small percentage of reservation families.

2. This association still troubles the Menominee as they observe drug and alcohol usage on the reservation to this day. Sylvia Wilber, the former Menominee director of the Maehnowseikiyah drug and alcohol treatment center on the reservation, so frustrated many members of the tribal leadership in 1995 by insisting that the whole Menominee community was in denial of the endemic nature of drug and alcohol abuse on the reservation that she was fired from her job as director. There is still a sense by some on the reservation that alcohol and drugs can make you, for at least a short while, more than you really are—that somehow these chemicals can put you into contact with the world of dreams and the spirits where power is given and taken away from individuals.

Chapter 7. A History of the Menominee Forest

1. The story of how the Court of Claims victory caused termination is told in chapter 5.

2. Some of the insect infestation techniques had to be abandoned after the foresters became more aware of the harmful side effects of insecticides after the publication of Rachel Carson's book *Silent Spring* (Carson 1970).

Bibliography

Abbey, Edward. 1977a. "Desert Places." In *The Journey Home*, 62–70. New York: E. P. Dutton.

———. 1977b. "The Second Rape of the West." In *The Journey Home*, 159–88. New York: E. P. Dutton.

Anderson, James E. 1994. *Public Policy-Making*. Geneva, Ill.: Houghton Mifflin.

André, Father. 1899. "Of the Mission of the Folle Avoine, Near the Bay Des Puants." In *Les Relations des Jesuites*, 58:272–81. Cleveland, Ohio: The Burrow Brothers Co.

Aplet, Gregory H.; Nels Johnson; Olson T. Jeffrey; and V. Alaric Sample, eds. 1993. *Defining Sustainable Forestry*. Washington, D.C.: Island Press.

Appenzeller, Tim. 1993. "Environment and the Economy." *Science* 260 (June): 1883.

Askinette, Steven. 1995. Conversation with the author. Keshena, Wis., February.

Ayer, Edward E. 1914. *Report on Menominee Indian Reservation*. Washington, D.C.: Department of the Interior.

Bartlett, Robert V. 1986. "Ecological Rationality: Reason and Environmental Policy." *Environment Ethics* 8 (Fall): 221–40.

Beck, David R. M. 1994. "Siege and Survival: Menominee Responses to an Encroaching World." Doctoral diss., University of Illinois at Chicago.

Berry, Thomas. 1988. *The Dream of the Earth*. San Francisco: Sierra Club Books.

Berry, Wendell. 1977. *The Unsettling of America Culture and Agriculture*. New York: Avon.

———. 1981a. *The Gift of Good Land*. San Francisco: North Point Press.

———. 1981b. *Recollected Essays, 1965-1980*. San Francisco: North Point Press.

———. 1994. "Conserving Forest Communities." *Another Turn of the Crank*, 25–45. Washington, D.C.: Counterpoint.

Bieder, Robert E. 1995. *Native American Communities in Wisconsin, 1600–1960*. Madison: University of Wisconsin Press.

Bloomfield, Leonard. 1928. *Menomini Texts*. New York: Publications of the American Ethnological Society, G. E. Strickert and Company.

Boulding, Kenneth. 1964. *The Meaning of the Twentieth Century*. New York: Harper and Row.

———. 1980. "The Economics of the Coming Spaceship Earth." In *Economics, Ecology, Ethics*, edited by Herman E. Daly, 253–63. New York: W. H. Freeman, 253–63.

———. 1991. "What Do We Want To Sustain?: Environmentalism and Human Evaluations." In *Ecological Economics*, edited by Robert Costanza, 22–31. New York: Columbia University Press.

Boyd, Theodore S. 1969. Quoted in *Antigo Daily Journal*, 1 November.

Brown, Lester R., and Pamela Shaw. 1982. *Six Steps to a Sustainable Society*. Washington, D.C.: Worldwatch Institute.

Burr, Gordon. 1995. Conversation with the author. Shawano, Wis., 26 December.

Callicott, Baird J. 1983. "Traditional American Indian and Traditional Western European Attitudes Toward Nature: An Overview." In *Environmental Philosophy*, 231–59. Edited by Robert Elliot and Arran Gare. Queensland, Australia: University of Queensland Press.

Carley, Michael, and Ian Christie. 1993. *Managing Sustainable Development*. Minneapolis: University of Minnesota Press.

Carson, Rachael. 1970. *Silent Spring*. New York: Fawcett.

Chapman, H. H. 1994. *The Menominee Indian Timber Case History*. Prepared for History 288: Menominee History, 1866–1950, David R. Wrone. Keshena, Wis.: College of the Menominee Nation Press.

Chamberlin, J. E. 1975. *The Harrowing of Eden*. New York: Seabury.

Cobb, John. 1980. "Ecology, Ethics, and Theology." In *Economics, Ecology, Ethics*, edited by Herman E. Daly, 162–76. New York: W. H. Freeman, 162–76.

Cokain, Charles. 1993. "Implementing Values-Based Education Study and Analysis of 'The Menominee Model.'" Master's thesis, Marian College-Fond du Lac., Wis.

Commoner, Barry. 1972. *The Closing Circle, Nature, Man and Technology*. New York: Bantam Books.

"Cooperative Agreement Between the U.S. Forest Service and the Department of the Interior." 1908. Washington, D.C.: Department of the Interior, Bureau of Indian Affairs records, 28 January.

Cope, Alfred. 1993. *A Mission to the Menominee: Alfred Cope's Green Bay Diary*. Edited by David R. Wrone from articles taken from the *Wisconsin Magazine of History*. Keshena, Wis.: College of the Menominee Nation.

Costanza, Robert. 1991. "Assuring Sustainability of Ecological Economic Systems." In *Ecological Economics*, edited by Robert Costanza, 331–43. New York: Columbia University Press.

———, ed. 1991. *Ecological Economics*. New York: Columbia University Press.

Costanza, Robert; Herman E. Daly; and Joy Bartholomew. 1991. "Goals, Agenda, and Policy Recommendations for Ecological Economics." In *Ecological Economics*, edited by Robert Costanza, 1–20. New York: Columbia University Press.

Crow, Thomas R.; Alan Haney; and Donald M. Waller. 1994. *Report on the Scientific Roundtable on Biological Diversity Convened by the Chequamegon and Nicolet National Forests*. St. Paul, Minn.: North Central Forest Experiment Station, Forest Service, U.S. Department of Agriculture.

Culbertson, John M. 1971. *Economic Development: An Ecological Approach*. New York: Alfred A. Knopf.

Daly, Herman E. 1977. *Steady-State Economics*. San Francisco: W. H. Freeman.

———, ed. 1980. *Economics, Ecology, Ethics*. New York: W. H. Freeman.

Daly, Herman E., and J. B. Cobb Jr. 1989. *For the Common Good: Redirecting the Economy Toward Community, the Environment, and a Sustainable Future*. Boston: Beacon Press.

Daly, Shirley. 1994. Conversation with the author. Keshena, Wis., 21 August.

Deloria, Vine. 1969. *Custer Died for Your Sins; An Indian Manifesto*. New York: Simon and Schuster.

Deloria, Vine. 1970. *We Talk, You Listen*. New York: Dell.

Devall, Bill, and George Sessions. 1985. *Deep Ecology*. Salt Lake City, Utah: Peregrine Smith.

Dovers, Stephen R., and John W. Handmer. 1993. "Contradictions in Sustainability." *Environmental Conservation* 20, no. 3 (Autumn): 217–22.

Dryzek, John S. 1987. *Rational Ecology*. New York: Basil Blackwell.

Einbender, LeGrand, and D. B. Wood. 1991. "Social Forestry in the Navajo Nation." *Journal of Forestry*, January, 12–18.

Elliot, Robert, and Arran Gare, eds. 1983. *Environmental Philosophy*. Queensland, Australia: University of Queensland Press.

Erdrich, Louise. 1988. *Tracks*. New York: Perennial.

Flader, S. L. 1983. *The Great Lakes Forest: An Environmental and Social History*. Minneapolis: University of Minnesota Press.

Fowler, Verna. 1994a. Conversation with the author. Keshena, Wis., 7 May.

———. 1994b. Conversation with the author. In an airplane between Detroit and Washington, D.C., 15 October.

———. 1995. "Preface–1995." In *Freedom with Reservation*, by the National Committee to Save the Menominee People and Forest, edited by Deborah Shames. Keshena, Wis.: College of the Menominee Nation Press.

Gale, Richard P., and Sheila M. Cordray. 1991. "What Should Forests Sustain? Eight Answers." *Journal of Forestry*, May, 31–36.

Gandhi, Mahatma. 1991. *The Essential Writings of Mahatma Gandhi*. Edited by Raghavan Iyer. Delhi: Oxford University Press.

Gore, Al. 1992. *Earth in the Balance*. Boston: Houghton Mifflin.

Gunn, Bruce. 1992. "Competruism: Ideology with a Sustainable Future." *Futures*, July/August, 559–74.

Haas, Peter M.; Marc A. Levy; and Edward A. Parson. 1992. "The Earth Summit, How Should We Judge UNCED's Success?" *Environment* 34, no. 8 (October): 7–32.

Hall, Brian P.; Bruce Taylor; Janet Kalven; and Larry S. Rosen. 1991. *Developing Human Values*. Fond du Lac, Wis.: International Values Institute of Marian College.

Hardin, Garrett. 1980. "The Tragedy of the Commons." In *Economics, Ecology, Ethics*, edited by Herman E. Daly, 100–114. New York: W. H. Freeman.

———. 1991. "Paramount Position in Ecological Economics." In *Ecological Economics*, edited by Robert Costanza, 45–57. New York: Columbia University Press.

Hodson, H. V. 1972. *The Diseconomics of Growth*. London: Earth Island.

Hoffman, Walter James. 1896. *The Menomini Indians*, 11–328. Fourteenth Annual Report of the Bureau of Ethnology, J. P. Powell, Director. Washington, D.C.: Government Printing Office.

House, Charles. 1959. In "Letters to the Editor." *Green Bay Press Gazette*, 19 August.

Huff, Paula, and Marshall Pecore. 1995. "Case Study: Menominee Tribal Enterprises (Menominee Reservation, Wisconsin, USA)." Madison, Wis.: Land Tenure Center, Institute for Environmental Studies.

Jakes, Pamela, and Jan Harms. 1995. *Report on the Socioeconomic Roundtable Convened by the Chequamegon and Nicolet National Forests*. St. Paul, Minn.: North Central Forest Experiment Station, Forest Service, U.S. Department of Agriculture.

Jefferson, Thomas. 1984. *Writings*. New York: Literary Classics of the United States.

Kapp, William K. 1983. *Social Costs, Economic Development, and Environmental Disruption*. Edited by John E. Ullmann. Lanham, Md.: University Press of America.

Keesing, Felix M. 1987. *The Menomini Indians of Wisconsin*. Madison: University of Wisconsin Press.

Kinney, J. P. 1945. "Report on Menominee Indian Forest." Washington, D.C.: Government Printing Office.

Kraft, Michael E. 1992. "Ecology and Political Theory: Broadening the Scope of Environmental Politics." *Policy Studies Journal* 20, no. 4:712–18.

Kraft, Michael E. 1996. *Environmental Policy and Politics*. New York: HarperCollins.

Krohelski, James T.; Phil A. Kammerer Jr.; and Terrence D. Conlon. 1994. *Water Resources of the Menominee Indian Reservation of Wisconsin*. Madison, Wis.: U.S. Geological Survey.

LaFollette, Robert. 1908. Speech before the United States Senate. Washington, D.C., 23 January.

———. 1960. *LaFollette's Autobiography, A Personal Narrative of Political Experience*. Madison: University of Wisconsin Press.

Landis, Scott. 1992. "Seventh Generation Forestry." *Harrowsmith Country Life*, November/December, 27–33

Leopold, Aldo. 1966. *A Sand County Almanac*. New York: Oxford University Press.

Lopez, Barry Holstun. 1981. *Desert Notes: Reflections in the Eye of a Raven*. New York: Avon.

Lorenz, Konrad. 1970. *King Solomon's Ring*. New York: Thomas Y. Crowell.

———. 1973. *Civilized Man's Eight Deadly Sins*. Translated by Marjorie Kerr Wilson. New York: Harcourt Brace Jovanovich.

Lurie, Nancy O. 1980. *Wisconsin Indians*. Madison, Wis.: State of Wisconsin Historical Society.

McDonough, Kathy. 1993. *Omaeqnomenew Masenahekan: Facts and Figures*. Keshena, Wis.: Menominee Indian Tribe of Wisconsin.

McLuhan, T. C. 1971. *Touch the Earth*. New York: Promontory Press.

Menominee County Social Services. 1992. Files of the Menominee County Social Services Department, Keshena, Wisconsin.

Menominee Forestry Department, Menominee Tribal Enterprises. 1970. *Forest Management Plan*. Keshena, Wis.: Menominee Tribal Enterprises.

————. 1995. *Menominee Tribal Enterprises Forest Management Plan, 1996–2005*. Keshena, Wis.: Menominee Tribal Enterprises.

Menominee Indian Tribe, Land Use Commission. 1983. *The Comprehensive Land Use and Natural Resource Plan*. Keshena, Wis.: Menominee Indian Tribe of Wisconsin.

Menominee Indian Tribe of Wisconsin. 1993. *Constitution of the Menominee Indian Tribe of Wisconsin*. Keshena, Wisconsin: Menominee Indian Tribe of Wisconsin.

————. 1994. *1994 Annual Report*. Keshena, Wis.: Menominee Indian Tribe of Wisconsin.

Menominee Indian Tribe of Wisconsin's Enrollment Office. 1994. Conversation with the author, January.

Menominee Tribal Enterprises. 1985. *Annual Report 1985*. Neopit, Wis.: Menominee Tribal Enterprises.

————. 1986. *Annual Report 1986*. Neopit, Wis.: Menominee Tribal Enterprises.

————. 1987. *Annual Report 1987*. Neopit, Wis.: Menominee Tribal Enterprises.

————. 1988. *Annual Report 1988*. Neopit, Wis.: Menominee Tribal Enterprises.

————. 1989. *Annual Report 1989*. Neopit, Wis.: Menominee Tribal Enterprises.

————. 1990. *Annual Report 1990*. Neopit, Wis.: Menominee Tribal Enterprises.

————. 1991. *Annual Report 1991*. Neopit, Wis.: Menominee Tribal Enterprises.

————. 1992. *Annual Report 1992* . Neopit, Wis.: Menominee Tribal Enterprises.

————. 1993. *Annual Report 1993* . Neopit, Wis.: Menominee Tribal Enterprises.

————. 1994. *Annual Report 1994*. Neopit, Wis.: Menominee Tribal Enterprises.

————. 1994. "Menominee Tribal Enterprises." brochure

Menominee Sustainable Development Institute. 1996. "Technical Brochure on the Menominee Forest." Rough draft. Keshena, Wis.: College of the Menominee Nation.

Milbrath, Lester W. 1989. *Envisioning a Sustainable Society: Learning Our Way Out*. Albany: State University of New York Press.

Miller, Glen. 1994. Conversation with the author. Keshena, Wis., 28 August.

————. 1995. "Menominee Indian Tribe of Wisconsin: Keepers of the Forest." Paper submitted to the President's Council on Sustainable Development at the request of the President's Council.

Mitchell, Robert Cameron. 1990. "Public Opinion and the Green Lobby: Poised for the 1990s?" *Environmental Policy in the 1990s*, edited by Norman J. Vig and Michael E. Kraft, 81–99. Washington, D.C.: Congressional Quarterly Press.

Mladenoff, David J., and John Pastor. 1993. "Sustainable Forest Ecosystems in the Northern Hardwood and Conifer Forest Region: Concepts and Management." In *Defining Sustainable Forestry*, edited by Gregory H. Aplet; Nels Johnson; Jeffrey T. Olson; and V. Alaric Sample, 145–80. Washington, D.C.: Island Press.

National Commission on the Environment. 1993. *Choosing a Sustainable Future.* Washington, D.C.: Island Press.

National Committee to Save the Menominee People and Forests. 1995. *Freedom With Reservation.* Edited by Deborah Shames. Keshena, Wis.: College of Menominee Nation Press.

Newman, James Gilbert. 1967. *The Menominee Forest of Wisconsin: A Case History in American Forest Management.* Ann Arbor, Mich.: University Microfilms, Inc.

Norgaard, R. B. 1989. "The Case for Methodological Pluralism." *Ecological Economics* 1:37–57.

Opuhls, William, and Stephen Boyan Jr. 1992. *Ecology and the Politics of Scarcity Revisited: The Unraveling of the American Dream.* New York: W. H. Freeman.

Oshkosh, Reginald. 1930. Remarks made at a General Council meeting. From minutes of the meeting of 9 August, Keshena, Wis.

Ourada, Patricia K. 1979. *The Menominee Indians.* Norman: University of Oklahoma Press.

Paehlke, Robert C. 1990. "Environmental Values and Democracy: The Challenge of the Next Century." In *Environmental Policy in the 1990s*, edited by Norman J. Vig and Michael E. Kraft, 349–67. Washington, D.C.: Congressional Quarterly Press.

Page, Talbot. 1991. "Sustainability and the Problem of Valuation." In *Ecological Economics*, edited by Robert Costanza, 58–74. New York: Columbia University Press.

Parson, Edward A., Peter M. Haas, and Marc A. Levy. 1992. "Summary of UNCED." *Environment*, October, 12–13.

Pecore, Marshall. 1992. "Menominee Sustained Yield Management." *Journal of Forestry* 90, no. 7 (July): 12–16.

———. 1994. Conversation with the author. Keshena, Wis., 15 June.

Peroff, Nicholas. 1982. *Menominee Drums.* Norman: University of Oklahoma Press.

Piaget, Jean. 1968. *Six Psychological Studies*. New York: Vintage.

Powell, John Wesley. 1962. *Report on the Lands of the Arid Region of the United States*. Cambridge: The Belknap Press of Harvard University Press.

Pubanz, Dan. 1995. "Landscape-Scale Planning and Decision Making Using Vegetative Habitat Type Classification." Paper presented at the Nineteenth Annual National Indian Timber Symposium. Yakima, Wash., 24–28 April.

Pyatskowit, Ruth N., and Thomas Madsen. 1992. "Case Study of Early Intervention and Pre-school Groups on the Menominee Indian Reservation." Keshena, Wis.: College of the Menominee Nation files.

Ray, Verne F. 1971. *The Menominee Tribe of Indians, 1940–1970*. Washington, D.C.: United States Court of Claims Docket no. 134-87, *Menominee Tribe et al. v United States of America*, Plaintiff's Exhibit no. R.1, 1 November.

Redclift, Michael. 1987. *Sustainable Development*. London: Methuen.

Reed, Gerard. 1986. "A Native American Environmental Ethic." In *Religion and the Environmental Crises*, edited by Eugene C. Hargrove, 25–37. Athens: University of Georgia Press.

Ritzenthaler, Robert E. 1992. "The Menominee Indian Sawmill: A Successful Community Project." *Wisconsin Archaeologist*, vol. 32, no. 2, 39–44.

Robbins, William G. 1982. *Lumberjacks and Legislators: Political Economy of the U.S. Lumber Industry, 1890–1941*. College Station: Texas A&M University Press.

Robertson, Melvin L. 1961. *Chronology of Events Relating to Termination of Federal Supervision of the Menominee Indian Reservation*. Minneapolis, Minn.: U.S. Department of the Interior, Bureau of Indian Affairs.

Robertson, W. A. 1993. "New Zealand's New Legislation for Sustainable Resource Management." *Land Use Policy* 10, no. 4 (October): 303–11.

Ruckleshaus, William D. 1989. "Toward a Sustainable World." *Scientific American*, September, 0036/9733, 166–70.

Salwasser, Hal; Douglas W. MacClery; and Thomas A. Snellgrove. 1993. "An Ecosystem Perspective on Sustainable Forestry and New Directons for the U.S. National Forest System." In *Defining Sustainable Forestry*, edited by Gregory H. Aplet, Nels Johnson, Jeffrey T. Olson, and V. Alaric Sample, 44–89. Washington, D.C.: Island Press.

Sax, Joseph L. 1993. "Property Rights and the Economy of Nature: Understanding *Lucas v. South Carolina Coastal Council*." *Stanford Law Review* 45, 1433, 153–75.

Shawano Leader. 1995. "Menominees Honored at UN Ceremony for Forest Practices." 21 April, 1.

Shinners, Hucovski & Company. 1993. Menominee Indian Tribe of Wisconsin Financial Statements as of September 30, 1993. Green Bay, Wis.

Shumacher, E. F. 1975. *Small Is Beautiful*. New York: Perennial.

Skinner, Alanson. 1913. "Social Life and Ceremonial Bundles of the Menomini Indians." In vol. 12, pt. 1, of the *Anthropological Papers of the American Museum of Natural History*. New York: The American Museum of Natural History.

———. 1921. Material Culture of the Menomini. *Indian Notes and Monograph Series*, edited by William H. Hodge. New York: Museum of the American Indian, Heye Foundation.

Snyder, Gary. 1990. *The Practice of the Wild*. New York: North Point Press.

Soil Survey Division, Wisconsin Geological and Natural History Survey. 1967. "Soil Survey Bulletin of Menominee County, Wisconsin." Madison: University of Wisconsin Press.

Spindler, George, and Louise Spindler. 1984. *Dreamers with Power*. Prospect Heights, Ill.: Waveland Press.

Stearns, Forest. 1987. "The Changing Forests of the Lake States." In *The Lakes States Forests*, edited by William E. Shands, 25–35. St. Paul, Minn.: The Lakes States Forestry Alliance.

Stegner, Wallace. 1954. *Beyond the Hundredth Meridian*. New York: Penguin Books.

Stivers, Robert L. 1976. *The Sustainable Economy, Ethics and Economic Growth*. Philadelphia: The Westminster Press.

Stone, Deborah A. 1988. *Policy Paradox and Political Reason*. Glenview, Ill.: Scott, Foresman.

Supreme Court of the United States, Menominee Tribe of Indians et al. v The United States of America. 1984. Supplemental Appendices G & H to Petition for A Writ of Certiorari to the United States Court of Appeals for the Federal Circuit. Washington, D.C.: Government Printing Office, May 25.

Temple, Stanley. 1992. "Old Issue, New Urgency?" Interview in *Wisconsin Environmental Dimension* 1, no. 1 (Spring): 1.

Tempus, Kent. 1995. "Gaming, Economy Aid Job Picture." *Shawano Leader*, 3 July, 1.

Thoreau, Henry David. 1946. *Walden*. New York: Dodd, Mead.

Tibbets, Holly Youngbear. 1992. "Unsaid at UNCED: Songs of the Miner's Canary and Other Laments of the Environment and Development." Paper presented at the Twenty-Seventh Annual Geographic Congress, Washington, D.C.

Tietenberg, Tom. 1994. *Environmental Economics and Policy*. New York: HarperCollins.

United States Geographical Survey. 1994. *Water Resources of the Menominee Indian Tribe of Wisconsin*. Madison, Wis.: U.S. Geological Survey.

United States of America. 1965. "Menominee Indian Timber Cutting Act." *Journal of the Wisconsin Indians Research Institute* 1, no. 1 (March): 105–6.

Waukau, Jerry. 1994. Conversation with the author. Keshena, Wis., 23 November.

Waukau, Larry. 1994. Conversation with the author. Neopit, Wis., 24 November.

———. 1995. Conversation with the author. Neopit, Wis., 3 June.

Whaley, Rick, and Walter Bresette. 1994. *Walleye Warriors*. Philadelphia: New Society Publishers.

Wisconsin Historical Collections. 1876. Vol. 7. Madison, Wis.: State Historical Society of Wisconsin.

———. 1898. Vol. 14. Madison, Wis.: State Historical Society of Wisconsin.

World Commission on Environment and Development. 1987. *Our Common Future*. Oxford: Oxford University Press.

Yaffee, Stephen Lewis. 1994. *The Wisdom of the Spotted Owl*. Washington, D.C.: Island Press.

Youth Intervention Project Survey. 1994. "Tapping into Teen Concerns: Perceptions and Behavior in Menominee Nation/County, Wisconsin." Great Lakes Intertribal Council and University of Wisconsin School of Pharmacy.

Index